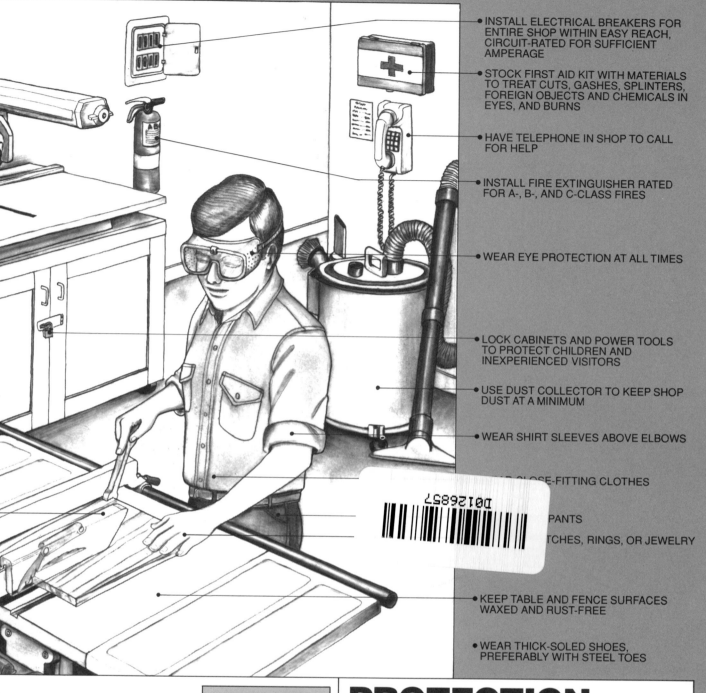

- INSTALL ELECTRICAL BREAKERS FOR ENTIRE SHOP WITHIN EASY REACH, CIRCUIT-RATED FOR SUFFICIENT AMPERAGE

- STOCK FIRST AID KIT WITH MATERIALS TO TREAT CUTS, GASHES, SPLINTERS, FOREIGN OBJECTS AND CHEMICALS IN EYES, AND BURNS

- HAVE TELEPHONE IN SHOP TO CALL FOR HELP

- INSTALL FIRE EXTINGUISHER RATED FOR A-, B-, AND C-CLASS FIRES

- WEAR EYE PROTECTION AT ALL TIMES

- LOCK CABINETS AND POWER TOOLS TO PROTECT CHILDREN AND INEXPERIENCED VISITORS

- USE DUST COLLECTOR TO KEEP SHOP DUST AT A MINIMUM

- WEAR SHIRT SLEEVES ABOVE ELBOWS

- WEAR CLOSE-FITTING CLOTHES

- PANTS

- WATCHES, RINGS, OR JEWELRY

- KEEP TABLE AND FENCE SURFACES WAXED AND RUST-FREE

- WEAR THICK-SOLED SHOES, PREFERABLY WITH STEEL TOES

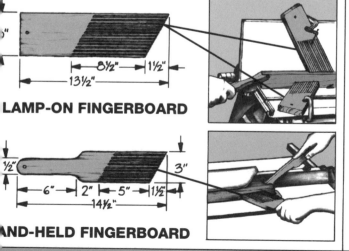

8½" 1½"
13½"

LAMP-ON FINGERBOARD

½"
3"
6" 2" 5" 1½"
14½"

AND-HELD FINGERBOARD

PROTECTION

WEAR FULL FACE SHIELD DURING LATHE TURNING, ROUTING, AND OTHER OPERATIONS THAT MAY THROW CHIPS

WEAR DUST MASK DURING SANDING AND SAWING

WEAR VAPOR MASK DURING FINISHING

WEAR SAFETY GLASSES OR GOGGLES AT ALL TIMES

WEAR RUBBER GLOVES FOR HANDLING DANGEROUS CHEMICALS

WEAR EAR PROTECTORS DURING ROUTING, PLANING, AND LONG, CONTINUOUS POWER TOOL OPERATION

THE WORKSHOP COMPANION®

USING THE DRILL PRESS

TECHNIQUES FOR BETTER WOODWORKING

by Nick Engler

Rodale Press
Emmaus, Pennsylvania

Printed in the United States of America on acid-free ∞, recycled ✿ paper

If you have any questions or comments concerning this book, please write:
Rodale Press
Book Readers' Service
33 East Minor Street
Emmaus, PA 18098

About the Author: Nick Engler is an experienced wood-worker, writer, teacher, and inventor. He worked as a luthier for many years, making traditional American musi-cal instruments before he founded *Hands On!* magazine. He has taught at the University of Cincinnati and gives wood-working seminars around the country. He contributes to woodworking magazines and designs tools for America's Best Tool Company. This is his forty-second book.

Series Editor: Kevin Ireland
Editors: Ken Burton
              Roger Yepsen
Copy Editor: Barbara Webb
Graphic Designer: Linda Watts
Illustrator: Mary Jane Favorite
Master Craftsman: Jim McCann
Photographer: Karen Callahan
Cover Photographer: Mitch Mandel
Proofreader: Hue Park
Indexer: Beverly Bremer
Interior and endpaper illustrations by Mary Jane Favorite
Produced by Bookworks, Inc., West Milton, Ohio

The author and editors who compiled this book have tried to make all the contents as accurate and as correct as possible. Plans, illustrations, photographs, and text have all been carefully checked and cross-checked. However, due to the variability of local conditions, construction materials, personal skill, and so on, neither the author nor Rodale Press assumes any re-sponsibility for any injuries suffered, or for damages or other losses incurred that result from the material presented herein. All instructions and plans should be care-fully studied and clearly understood be-fore beginning construction.

Special Thanks to:

Delta International Machinery
   Corporation
Pittsburgh, Pennsylvania

Shopsmith, Inc.
Dayton, Ohio

Wertz Hardware
West Milton, Ohio

Library of Congress Cataloging-in-Publication Data

Engler, Nick.
      Using the drill press/by Nick Engler
            p.   cm. — (The Workshop companion)
      Includes index.
      ISBN 0–87596–721–3 hardcover
      1. Drill presses — Handbooks, manuals, etc. I. Title.
      II. Series: Engler, Nick. Workshop companion.
TJ1260.E53   1995
684 '.083 — dc20   95–8114

   4 6 8 10 9 7 5 3   hardcover

CONTENTS

TECHNIQUES

PROJECTS

TECHNIQUES

1

DRILL PRESSES, BITS, AND ACCESSORIES

From prehistoric times, craftsmen understood that to drill a hole, they had to *feed* (apply pressure to) the bit. At first, they applied this pressure with their arms or chest, pushing the bit through the wood as they spun it with a bow string. Later, they began to use rocks and other heavy objects to press the bit into the wood. By the fifteenth century, woodworkers were boring holes with a *beam press* — a large wooden lever that pressed against the head of a brace and bit. In the late eighteenth and nineteenth centuries, mechanical *smith's drills* appeared, with simple foot treadles or screws to feed the bit. Over several decades, these became more complex and evolved into *bench drills* that used ingenious ratchets and gear-driven feeds. Tool manufacturers married the bench drill to an electric motor early in the twentieth century, creating the first modern *drill press*.

The drill press is designed as a metalworking tool, but it has found its way into woodworking shops because of the ease and accuracy it offers. Today, many craftsmen consider the drill press an essential tool for fine woodworking — you have a difficult time drilling a hole at a precise angle or depth without one. And its role has expanded. In addition to drilling and boring, you can also use it for some other workshop tasks with the proper fixtures and accessories. These include mortising, sanding, turning, routing, shaping, grinding, and sharpening.

A 1931 Model 620 Drill Press, reprinted with permission and courtesy of Delta International Machinery Corporation.

2

CHOOSING A DRILL PRESS
HOW A DRILL PRESS WORKS

Drill presses incorporate three major assemblies — the *head,* which holds, turns, and feeds the drill bit; the *table,* which supports the work; and the *column,* which supports both the head and the table. (*SEE FIGURES 1-1 AND 1-2.*)

The head holds the *motor,* which turns a *spindle* in a movable *quill.* Attached to the bottom end of the spindle is a *chuck* to hold drill bits. Typically, the power from the motor is transferred to the spindle through a *speed changer* (such as step pulleys or variable-diameter pulleys), letting you adjust the speed at which the drill bit turns. To feed or retract the bit, move the quill ver-

tically with the *quill feed* levers. Use the *depth stop* to limit the quill travel and the *quill lock* to secure the quill in one position.

The table rests just under the spindle. A *tilt adjustment* changes the angle of the work surface from side to side, and a *height adjustor* raises or lowers the table. Use the *table lock* to secure the table at a particular height. You can also move the table from side to side by loosening the table lock and rotating the table partway around the column.

The head is fastened to the top of the column, while the table slides up and down it. The bottom of

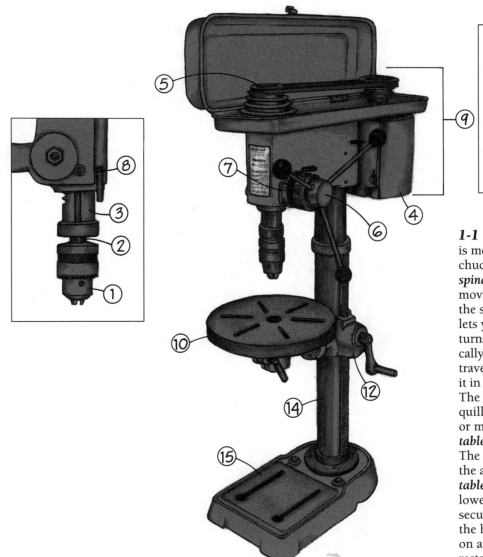

1-1 On a drill press, a drill bit is mounted in the *chuck* (1). The chuck is fastened to the end of a *spindle* (2), which rotates inside a movable *quill* (3). A *motor* (4) turns the spindle, and a *speed changer* (5) lets you adjust the speed at which it turns. You can move the quill vertically with a *quill feed* (6), limit its travel with a *depth stop* (7), or secure it in one position with a *quill lock* (8). The motor, drive train, quill, and quill adjustments are all encased in or mounted on the *head* (9). The *table* (10) rests just under the head. The *table tilt adjustment* (11) changes the angle of the work surface, the *table height adjustor* (12) raises or lowers it, and the *table lock* (13) secures the table in position. Both the head and the table are mounted on a *column* (14), and the column rests in a *base* (15).

the column is mounted in a large, flat *base*. The base rests on the floor, a workbench, or a stand. **Note:** On many drill presses, the base doubles as another table and is sometimes referred to as the *lower table*. It's ground flat to support work that is too tall to rest on the main table.

Not all drill presses have these components, of course. Small, inexpensive presses have simpler drive trains. They may lack quill locks and table height adjustors. Many older presses also have no table height adjustor — you must loosen the table lock and slide the table along the column to change the table height. Industrial presses often use tapered sockets to mount bits, rather than adjustable chucks. (*See Figure 1-3.*)

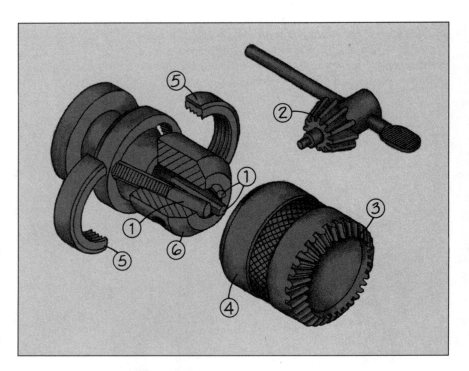

1-2 A drill press chuck adjusts to hold drill bits of different diameters. Most chucks have three movable *jaws* (1). When you use a geared *chuck key* (2) to turn the *ring gear* (3) and the outer *sleeve* (4) clockwise, a *split nut* (5) inside the chuck pushes the jaws through angled holes in the *chuck body* (6), bringing them closer together to clamp the bit. Turning the ring gear counterclockwise pulls them apart and releases the bit.

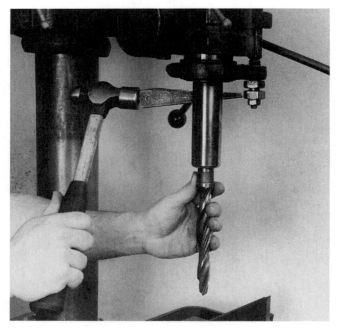

1-3 Many industrial drill presses are fitted with sockets rather than chucks. These sockets are machined with a standard *locking* taper to hold drill bits with tapered shanks. (Locking tapers have a slope of less than 7 degrees. There are several standard systems, the most common of which are *Morse* tapers.)The advantage of this system is that the bits are easier to mount and release. To mount a bit, simply slide it into the socket quickly. To release it, use a wedge-shaped *drift pin,* as shown.

Drill presses also have different capacities. The *chuck capacity* — how wide the jaws will open — determines the largest drill bit that you can use on that machine. The *throat capacity* — the horizontal distance from the chuck to the column — determines how far you can drill a hole from the edge of a work-piece. The *stroke* — the vertical distance that the quill extends from the head — tells you how deep you can make a hole in a single pass. Finally, the *work capacity* — the vertical distance between the chuck and the base — determines the largest workpiece you can drill. (*See Figure 1-4.*)

FOR YOUR INFORMATION

Manufacturers often double the throat capac-ity of a press when advertising their machines, telling you their "16-inch drill press" will drill to the center of a 16-inch-wide board — meaning it has an 8-inch throat capacity. This number is sometimes referred to as the *swing*. It's equal to the diameter of the circle the chuck would describe were you to swing the head around the column, minus the diameter of the column.

TYPES OF DRILL PRESSES

There are three different types of drill presses, each of which offers its own advantages.

■ The *floor drill press* rests on the floor, standing between 52 and 76 inches high. It has a much larger work capacity than other types. (*See Figure 1-5.*)

■ The *bench drill press* may have exactly the same head as a floor model, but it stands only 22 to 46 inches high and rests on a workbench. It has a smaller work capacity, but it's more compact. (*See Figure 1-6.*)

■ The *radial drill press* is about the same size as a bench model, but the throat adjusts to between 6 and 17 inches, typically. The head rotates to change the angle at which the quill feeds the drill bit. (*See Figure 1-7.*)

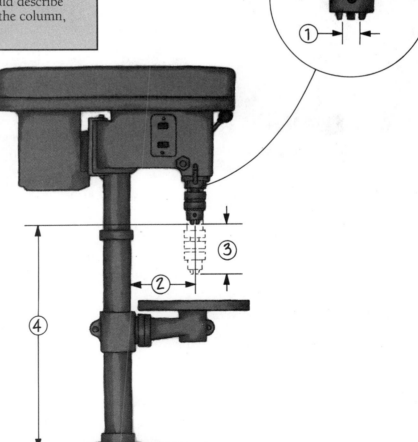

1-4 Drill presses have different types of capacities that determine the sizes of the bits you can mount and what work you can do with them. The **chuck capacity** (1) is the largest drill bit shank that particular chuck will hold. The **throat capacity** (2) determines how far from the edge of a board you can drill. The **stroke** (3) of the quill tells you how deep you can plunge a bit into a board. The **work capacity** (4) determines the longest, widest, and tallest work-pieces you can drill.

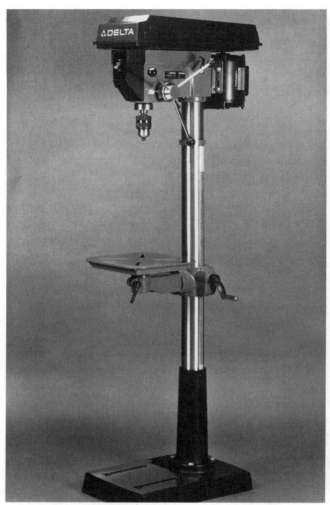

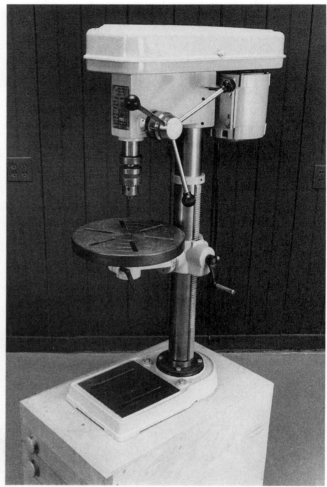

1-5 The *floor drill press* usually stands between 52 and 76 inches tall, and the base rests on the floor. This offers a large work capacity — if you need to, you can swing the table out of the way, rest a big workpiece on the base, and drill it. *Photo courtesy of Delta International Machinery Corporation.*

1-6 The *bench drill press* is typi- cally 22 to 46 inches high, and the base rests on a workbench or tool stand. Because of this, it has a much smaller work capacity than a floor model. However, its capacity is adequate for most woodworking operations, and you can use the space under the drill press to store bits and accessories.

FEATURES TO CONSIDER

When you select a drill press for your shop, consider these features:

CAPACITIES

When comparing chucks, throats, strokes, and work capacity, keep in mind the general rule that the larger, the better. But there are other considerations.

■ *Chucks* for presses come in ⅜-, ½-, and ⅝-inch sizes. Since most drills have shanks that are ½ inch in diameter or smaller, select a machine with at least a ½-inch chuck; ⅝ inch gives you a little extra room. Make sure the motor has adequate power to drive large bits. **Note:** If you need to mount extremely small bits, purchase a special *zero chuck* (also called a wire drill chuck). These chucks are attached to a shaft and can be secured in the standard chuck.

■ *Throats* on bench and floor models may be as small as 4 inches or as large as 10 inches. You can perform most drilling and boring operations with a

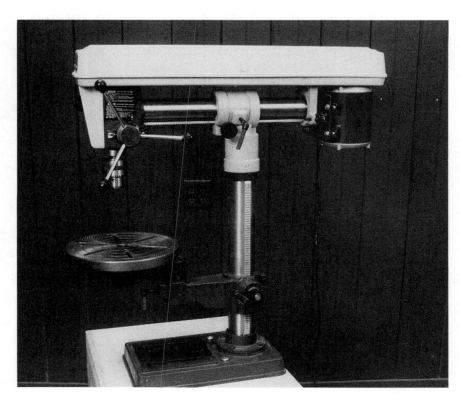

1-7 A *radial drill press* is about the same size as a bench model with a similar work capacity, but it has a much larger throat capacity (up to 17 inches) than either a bench model or a floor model. Typically, the head travels forward and back on a horizontal *ram column*, making the throat adjustable. The head and the ram column rotate right and left, letting you change the angle at which you feed the bit. As you might expect, these features make a radial drill press more versatile — and more expensive — than other types. **Note:** The extra moving parts on a radial drill press also make it more susceptible to manufacturing problems. If you purchase one, buy a reputable brand and carefully check its alignment. In particular, check that the chuck holds the bit square to the table no matter how you adjust the throat capacity.

machine that has a 7- or 8-inch throat. Depending on the projects you build, however, you may need something larger. If you do need a larger model, consider a radial drill press.

■ *Strokes* vary between 2 and 6 inches. Generally, a drill press with a 3- or 4-inch stroke will serve you well; drill bits are rarely longer than that. But again, the type of work you do may necessitate a longer stroke.

■ *Work capacity* is, for many woodworkers, the least important of these four variables. There is enough room under the head of an average bench drill press for all but a handful of drilling operations, and you can use a portable drill to fill in for most of those. However, beware of machines with extremely short columns: They may not offer enough capacity to use long bits or drill large workpieces.

MOTOR

To drill large holes (over 1 inch in diameter) in hardwood or to use hole saws, the motor should deliver at least ⅓ horsepower; ½ horsepower is better. Look for an *induction* motor that runs at a constant 1,725 rpm. These deliver more torque, run quieter, and hold up better during long sessions than *universal* motors. (Universal motors are better suited for portable power tools.) Also avoid 1,140-rpm induction motors — these low-speed motors are intended for metalworking tools.

SPEEDS

To cut different materials and sizes of holes, you need a range of speeds between 250 and 3,000 rpm, approximately. Because induction motors run at a constant speed, most manufacturers install mechanical speed changers in their drill presses — step pulleys or variable-diameter pulleys. It's easier to change the speed of a press with variable-diameter pulleys, but they are much more expensive. Think twice before you purchase a press with an electronic speed changer; these always have a universal motor.

QUILL

Look for a quill assembly with two bearings inside the quill. These help the spindle to run true. Quills with a single bearing often have excessive side-to-side play. (*See Figure 1-8.*) If you plan to perform operations like sanding or routing in which you must feed the work into the drill press from the side, choose a machine with a quill lock that pinches the quill in the casting and prevents it from moving sideways. (*See Figure 1-9.*)

TABLE

The larger the table, the more support it will provide for the work. More important than a large table, however, is one that can be easily adjusted. Look for a machine with a rack-and-pinion table height adjustment that lets you crank the table up and down.

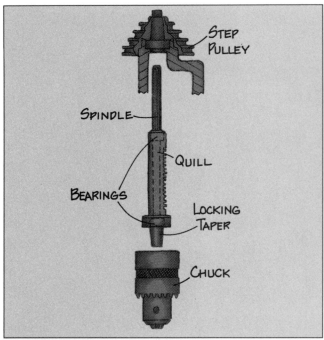

1-8 The quill is the heart of your drill press, and the better it's made, the easier you'll find it to drill accurately. Look for a quill with two bearings, top and bottom. These reduce side-to-side play and help the spindle run true.

1-9 A *quill lock* (1) should pinch the quill in the head casting to limit horizontal play as well as hold the quill's vertical position. The casting on the press shown is split (2) so the lock draws the two halves together around the quill. Other presses may use cams or screws to lock the quill down. Locks that simply prevent the pinion gear on the quill feed from turning are worthless; they don't eliminate horizontal *or* vertical play.

A BIT OF ADVICE

Some of the best drill presses available are multipurpose tools such as a Shopsmith or Total Shop. These have relatively large capacities, powerful induction motors, variable-diameter pulleys to change speeds, two-bearing quills with locks that pinch the quills in the head castings, large tables with fences, and the ability to drill horizontally as well as vertically. They only lack table height adjustment.

EXCEEDING THE CAPACITIES

No matter how large the throat and the work capacity of your drill press, sooner or later you'll need to drill a workpiece that exceeds them.

When faced with this dilemma, use a portable drill instead. Two attachments for hand-held electric drills provide drill press accuracy.

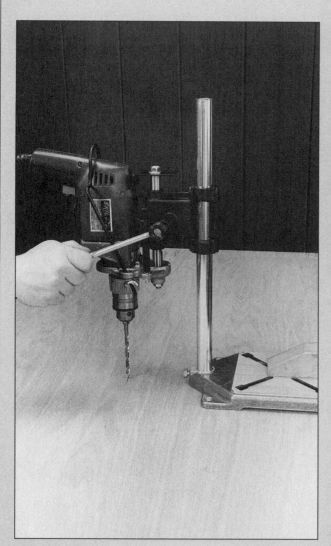

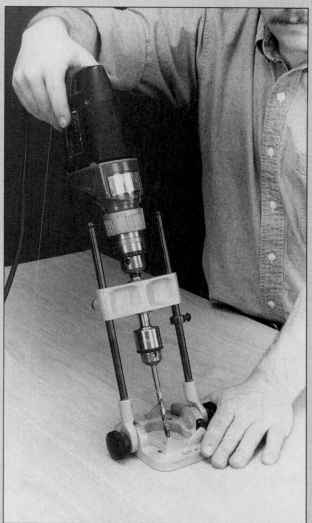

1 **A *drill stand* converts your** portable drill into a small drill press that you can set up right *on* the work. Turn the base around so it won't interfere with the bit, and clamp it to the work. In some cases, you may have to attach the base to a large scrap, then secure this scrap to the work.

2 **If you must drill precisely** *angled* holes with a portable drill, use a *drill guide*. This accessory doesn't have a feed; it simply guides the bit between two parallel rails. As with a drill stand, you can either fasten the base of the guide directly to the work or mount it on a separate board and clamp this board to the work.

DRILL BITS

GENERAL-PURPOSE BITS

There are hundreds of types of drill bits, most of them designed for specific industrial applications. Only a handful are made for general drilling operations — twist bits, brad-point bits, spade bits, boring bits, Forstner bits, and multi-spur bits. These are the bits most often used in woodworking shops.

Every bit has three parts: a shank, a body, and a tip. The *shank* mounts in the chuck, the *body* extends out from it, and the *tip* does the cutting. The tip can be further subdivided. It has one or more cutting edges, called *lips,* which cut the wood fibers and create a hole. Some drill bit tips have additional cutting edges or *spurs* to shave the circumference of the hole. Most have a *lead point* to help start the bit and keep it running true. (*SEE FIGURE 1-10.*)

High-speed twist bits are the best all-around multi-purpose choice. These are designed to drill metal, but they also work well in wood and plastic. Several lengths and hundreds of diameters are available. (*SEE FIGURES 1-11 AND 1-12.*) You can purchase them in fractional sizes (given in 1/64-inch increments) or decimal sizes (usually given in the nearest .001-inch increments). Decimal sizes below .230 inch are numbered to correspond to American Wire Gauge sizes from #1 (the largest) to #80 (the smallest). Sizes above .230 inch are lettered from A (the smallest) to Z (the largest). The chart "Twist Bit Sizes" on page 15 shows all the commonly available bits. **Note:** Twist bits also come in metric sizes, but these are not commonly used in woodworking in the United States.

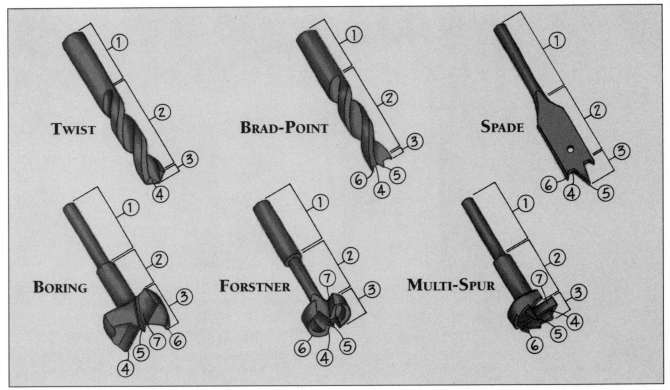

1-10 **There are six types of drill** bits commonly used in woodworking for general drilling and boring — *twist, brad-point, spade, boring, Forstner, and multi-spur bits.* Each bit has a **shank** (1) that mounts in the chuck, while the **body** (2) extends out in front of it. At the **tip** (3), the *lips* (4) cut the wood, creating the hole. Most woodworking bits also have a **lead point** (5) at the tip to help start the bit and keep it running true. Some bits, like the brad-point, Forstner, multi-spur, and boring bits, have *spurs* (6) to score the circumference of the hole out ahead of the tip. Twist bits and brad-point bits have spiral grooves or *flutes* (6) that lift the wood chips out of the hole. Others, such as the Forstner, multi-spur, and boring bits, have angled surfaces or *lifters* (7) to evacuate the chips.

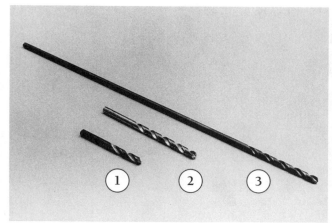

1-11 Twist bits are commonly
available in three different lengths.
The shortest are *stub* bits (1). These
are typically used in industrial applications or as pilot drills for hole saws
and fly cutters. The most commonly
available lengths are *jobber* bits (2).
These vary between 2 and 4 inches
long, approximately, depending on
their diameter. (The larger the diameter, the longer the drill.) The longest
are *extended* bits (3), measuring
between 10 and 12 inches long,
which are useful for drilling deep
holes. Because they are so often used
to drill holes through walls for
wiring, they are sometimes referred
to as *installer* bits.

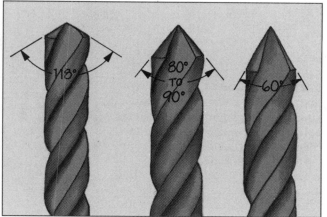

1-12 As they come from the
manufacturer, the tips of twist bits
are typically sharpened at 118
degrees, the optimum angle for
drilling metal. This angle also does a
fair job in wood and plastic, although
smaller angles will do better. The
best angles for drilling wood are
between 80 and 90 degrees, while the
best for plastic is 60 degrees. If you
regularly use twist bits to drill soft
materials, you'll get better results if
you regrind the tips (or have them
reground) at the appropriate angle.

Most woodworkers prefer *brad-point bits* (also called
machine spur bits) for general drilling operations in
wood. The lead point makes these bits easy to align
with layout marks, while spurs cut a clean entrance
hole. The bits also do well in plastic but cannot be
used to drill metal. Brad-point bits are commonly
available in fractional sizes from 1/16 inch to 1¼ inches
in diameter, and in metric sizes from 3 millimeters to
18 millimeters.

Spade bits bore quickly but leave a rough hole. The
lips of spade blades are miniature scraper blades that
scrape the wood away rather than shave it. They do

not evacuate wood chips — when boring deep holes,
you must frequently retract the bit and remove the
waste. However, these bits have distinct advantages.
They are excellent for boring end grain, and you can
easily regrind the profiles to make holes of odd sizes
and shapes. (*SEE FIGURE 1-13.*) They come in fractional
sizes from ¼ inch to 1½ inches.

Like spade bits, *boring bits* scrape the wood away
rather than cut it. However, they also have spurs to
scribe the hole as they scrape. This arrangement
allows them to cut just as quickly as spade bits, while
leaving a smoother hole. Unfortunately, they are more
expensive than spade bits and not as readily available.
In the United States, craftsmen use them primarily to
install hardware. They are available in metric sizes
from 15 millimeters to 40 millimeters in diameter.
You can also find a few fractional sizes.

Forstner bits are designed to cut flat-bottom holes
with smooth sides. (*SEE FIGURE 1-14.*) They have a tiny
lead point and long, semicircular spurs or *cutting rims*

TRY THIS TRICK

If you need brad-point bits in numbered or lettered sizes, have a good sharpener regrind the tips
on a set of numbered or lettered twist bits for you.

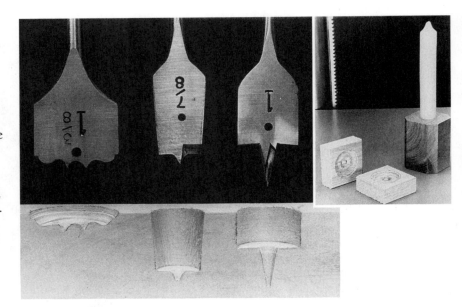

1-13 If you need to drill odd-size or odd-shaped holes, modify a spade bit. Select a bit slightly larger than the hole you want to drill, and grind the edges to reduce the diameter or change the profile. Sharpen the ground edges with a file, as described in "Sharpening Drill Bits" on page 44. The hole on the right was made with an ordinary spade bit. The side edges of the bit in the middle were ground to make a tapered candle hole, while the lips of the bit on the left were shaped to cut a decorative rosette.

1-14 A Forstner bit cuts a hole with a flat bottom and extremely smooth sides, as shown on the left. This is your best choice for making decorative holes or circular recesses in which the sides or the bottom of the recess will be seen. The hole on the right was drilled with a brad-point bit. As you can see, it's slightly rougher and does not have a flat bottom.

that slice the wood fibers at the circumference of the hole just ahead of the lifters. They come in fractional sizes from ¼ inch to 3 inches in diameter, and a few metric sizes are available for installing European cabinet hardware. The smaller sizes (under 1 inch) can also be used to cut plastic; the larger sizes become increasingly difficult to control and may melt the plastic at the cutting rim unless your drill press can be set to run at extremely low speeds.

Multi-spur bits are similar to Forstner bits, but they have a longer lead point and sawtooth spurs arranged around the cutting rims. They bore extremely smooth holes, although not quite as smooth as Forstner bits do. However, they are much easier to control when drilling large-diameter holes. Many craftsmen use

Forstner bits for holes up to 1½ inches in diameter, then switch to multi-spur bits for larger holes. They come in fractional sizes from ⅜ inch to 4 inches in diameter.

A SAFETY REMINDER

Do not use *auger bits* with lead screws (rather than lead points) on a drill press. These self-feeding bits are designed to be used with braces and hand drills. On a drill press, they pull the bit into the wood too quickly and the machine stalls.

SPECIAL-PURPOSE BITS AND CUTTERS

There are also several specialized bits and cutters that are commonly used with the drill press. (*SEE FIGURE 1-15.*)

Hole saws (also called *crown saws*) have a circular blade. A pilot bit helps to center the blade and keep it steady as you start the hole. They are useful for drilling large, rough holes and are available in fractional sizes up to 6 inches in diameter.

Fly cutters have a single cutting edge mounted on the end of an adjustable arm. By sliding the arm through the body, you can change the distance of the cutter from the center or *axis* of the fly cutter. This, in turn, changes the diameter of the circle you cut, letting you make holes up to 6 inches in diameter. A pilot bit holds the cutter steady as the arm swings around the body. Fly cutters are designed to cut holes in sheet metals, but they will also work with wood and plastic.

Countersinks and *center reamers* cut tapered holes. They can be used for centering lathe work, removing chips or burrs around the edges of holes, or making tapered recesses for flathead screws and bolts.

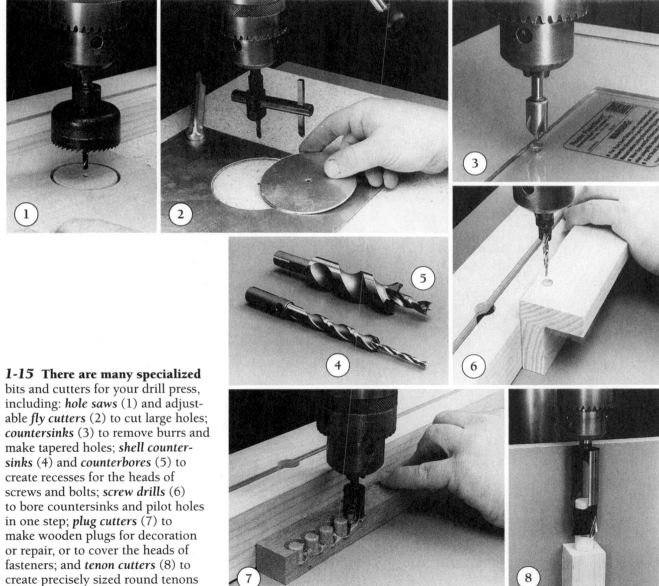

1-15 There are many specialized bits and cutters for your drill press, including: *hole saws* (1) and adjustable *fly cutters* (2) to cut large holes; *countersinks* (3) to remove burrs and make tapered holes; *shell countersinks* (4) and *counterbores* (5) to create recesses for the heads of screws and bolts; *screw drills* (6) to bore countersinks and pilot holes in one step; *plug cutters* (7) to make wooden plugs for decoration or repair, or to cover the heads of fasteners; and *tenon cutters* (8) to create precisely sized round tenons with flat shoulders.

Shell countersinks and *counterbores* also cut recesses for fasteners. Both cutters slip over twist bits or brad-point bits and lock in place with set screws, allowing you to bore a pilot hole and a recess in one step. A shell countersink creates tapered recesses for flathead hardware, while a counterbore makes recesses with straight sides and flat bottoms for other fasteners. (*See Figure 1-16.*)

Screw drills are matched shell countersinks and drill bits that are sold as sets. The shell often includes a narrow counterbore to make a shaft hole and a stop collar to control the depth of the cut. (*See Figure 1-17.*) The components are sized to correspond to common screw sizes. For example, a 5/32-inch bit and a 3/8-inch

shell countersink comprise a screw drill for a #8 flathead wood screw.

Plug cutters cut wooden plugs to fit holes and counterbores. These may be used to decorate a project, repair a wooden surface, or cover the heads of screws and other fasteners. Some are designed to cut straight plugs with a constant diameter; others cut tapered plugs for a tighter fit. Generally, the tapered plugs are less visible when installed.

Tenon cutters are similar to plug cutters, but they are designed to cut a long, round tenon with a flat shoulder. They can be used with both square and round stock.

1-16 *Shell countersinks* and *counterbores* slip over ordinary drill bits to produce recessed pilot holes for fasteners. A countersink makes a recess with tapered sides to hold a flathead screw or bolt. A counterbore creates a recess with straight sides and a flat bottom for other hardware.

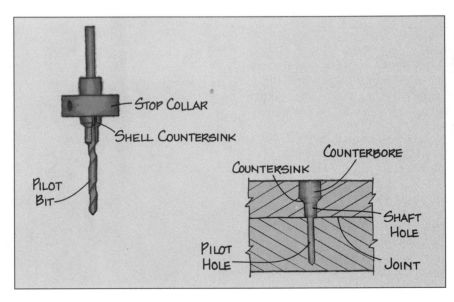

1-17 *A screw drill* often consists of a *pilot bit*, a *shell countersink*, and a *stop collar*. This combination enables you to drill a *pilot hole*, *countersink*, and *counterbore* to a precise depth all in one step. Sometimes the shell countersink is shaped to make a *shaft hole* for the unthreaded portion of the screw.

TWIST BIT SIZES

NUMBERED BITS							
Number	**Decimal Equivalent**	**Number**	**Decimal Equivalent**	**Number**	**Decimal Equivalent**	**Number**	**Decimal Equivalent**
80	.014″	60	.040″	40	.098″	20	.161″
79	.015″	59	.041″	39	.100″	19	.166″
78	.016″	58	.042″	38	.102″	18	.170″
77	.018″	57	.043″	37	.104″	17	.173″
76	.020″	56	.047″	36	.106″	16	.177″
75	.021″	55	.052″	35	.110″	15	.180″
74	.023″	54	.055″	34	.111″	14	.182″
73	.024″	53	.060″	33	.113″	13	.185″
72	.025″	52	.064″	32	.116″	12	.189″
71	.026″	51	.067″	31	.120″	11	.191″
70	.028″	50	.070″	30	.129″	10	.194″
69	.029″	49	.073″	29	.136″	9	.196″
68	.031″	48	.076″	28	.141″	8	.199″
67	.032″	47	.079″	27	.144″	7	.201″
66	.033″	46	.081″	26	.147″	6	.204″
65	.035″	45	.082″	25	.150″	5	.206″
64	.036″	44	.086″	24	.152″	4	.209″
63	.037″	43	.089″	23	.154″	3	.213″
62	.038″	42	.094″	22	.157″	2	.221″
61	.039″	41	.096″	21	.159″	1	.228″

FRACTIONAL BITS				LETTERED BITS			
Fraction	**Decimal Equivalent**	**Fraction**	**Decimal Equivalent**	**Letter**	**Decimal Equivalent**	**Letter**	**Decimal Equivalent**
1/64	.016″	17/64	.266″	A	.234″	N	.302″
1/32	.031″	9/32	.281″	B	.238″	O	.316″
3/64	.047″	19/64	.297″	C	.242″	P	.323″
1/16	.062″	5/16	.313″	D	.246″	Q	.332″
5/64	.078″	21/64	.328″	E	.250″	R	.339″
3/32	.094″	11/32	.344″	F	.257″	S	.348″
7/64	.109″	23/64	.359″	G	.261″	T	.358″
1/8	.125″	3/8	.375″	H	.266″	U	.368″
9/64	.141″	25/64	.391″	I	.272″	V	.377″
5/32	.156″	13/32	.406″	J	.277″	W	.386″
11/64	.172″	27/64	.422″	K	.281″	X	.397″
3/16	.188″	7/16	.438″	L	.290″	Y	.404″
13/64	.203″	29/64	.453″	M	.295″	Z	.413″
7/32	.219″	15/32	.469″				
15/64	.234″	31/64	.484″				
1/4	.250″	1/2	.500″				

DRILLING ACCESSORIES

In addition to bits and cutters, there are many accessories that can be used with drill presses to help control the bit, provide storage, or hold and position workpieces. Still other accessories allow you to use a drill press for woodworking operations such as planing, sanding, mortising, and grinding.

■ *Stop collars* fasten to twist bits or brad-point bits to limit the depth of the hole. (*SEE FIGURE 1-18.*)

■ A *drill press vise* fastens to the table and holds the work steady while you drill it. (*SEE FIGURE 1-19.*)

■ A *drill press clamp* secures the work to the table to hold it steady. (*SEE FIGURE 1-20.*)

■ A *surfacer* or *rotary planer* mounts in the drill chuck and lets you plane wood by passing the board under the accessory. (*SEE FIGURE 1-21.*)

■ A *mortising attachment* enables you to bore square holes and mortises with a drill press. (*SEE FIGURE 1-22.*)

■ *Drum sanders* and other sanding accessories let you perform a wide range of sanding operations on a drill press. (*SEE FIGURE 1-23.*)

■ *Rotary rasps* mount in the chuck and scrape away the wood to create curves and contours. (*SEE FIGURE 1-24.*)

1-18 A *stop collar* **fastens around** the circumference of the bit to limit its travel. This lets you bore holes to a precise depth. You can use the press's depth stop to do the same thing, of course, but stop collars don't have to be reset whenever you change the table height or the stock thickness.

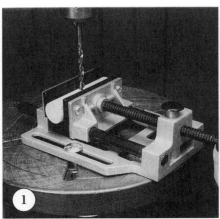

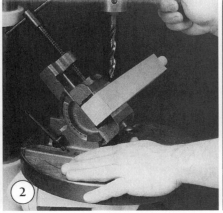

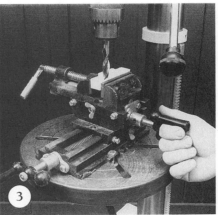

1-19 **There are several types of** vises made to attach directly to a drill press table. A *drill press vise* (1) clamps the work between two jaws and holds it steady as you work. An *angle vise* (2) holds the work and lets you adjust its angle up to 45 degrees. A *cross vise* (3) holds the work and lets you move it in two dimensions, front to back and side to side, making it possible to do simple milling and routing operations.

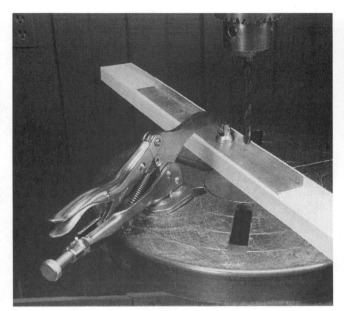

1-20 A *drill press clamp* fastens to the drill press and holds the work down on the table while you drill it. This particular clamp operates like locking-jaw pliers.

1-21 A *rotary planer* or *surfacer* mounts in a drill press chuck. As you pass wood beneath the planer, it shaves the surface. It doesn't have a large capacity — most rotary planers are no more than 3 inches in diameter — but you can make several passes to plane a fairly wide board. This accessory is particularly useful when you must plane figured wood — the hooked cutters on the bottom of the rotating disc are less likely to chip or tear the wood grain than the knives of a traditional planer.

1-22 A *mortising attachment* converts your drill press to a mortiser, letting you bore square holes. A square, hollow chisel fastens to the lower end of the quill, and a bit mounts in the chuck and extends down through the hollow chisel. A hold-down anchors to the table or the fence, depending on the design. Chisels and bits are available in sizes from $\frac{1}{4}$ to $\frac{5}{8}$ inch square.

■ *Grinding cups, wheels,* and *drums* let you use the drill press to grind and sharpen drill bits, chisels, knives, and other cutting tools. (*SEE FIGURE 1-25.*)

■ *Wire wheels* are useful for cleaning dirty or corroded metal parts. They can also be used to make wood look old and weathered. (*SEE FIGURE 1-26.*)

■ A swing-out *storage table* mounts to the drill press column to hold bits, keys, and other small drilling accessories within easy reach. (*SEE FIGURE 1-27.*)

■ An *auxiliary table* mounts on a standard drill press table and extends the work surface, helping to support large workpieces. (*SEE FIGURE 1-28.*)

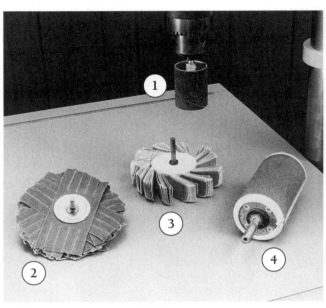

1-23 There are several different sanding attachments you can mount on a drill press to perform a variety of sanding operations. A *drum sander* (1) enables you to smooth edges and thickness-sand small parts. *Flutter sheets* (2) and *flap sanders* (3) sand curves and contours. A *pneumatic sander* (4) conforms to three-dimensional surfaces and blends hard edges.

1-24 Like a cabinetmaker's rasp, *rotary rasps* scrape away the wood surface, letting you create shapes and contours. They are generally used in portable drills and flexible-shaft tools, but they can also be used in drill presses. Select the rasp shape best suited for the contour you want to make, mount it in the chuck, and hold the workpiece lightly against it.

1-25 Shank-mounted *grinding* cups, wheels, and *drums* can all be fastened in the chuck for grinding and sharpening tasks. In many cases, a drill press is actually a better tool for these tasks than most grinders because it can be adjusted to turn at a low speed. The slower the grinding tool turns, the less friction it generates and the cooler it runs. This, in turn, helps keep the abrasives from overheating the metal and altering the temper.

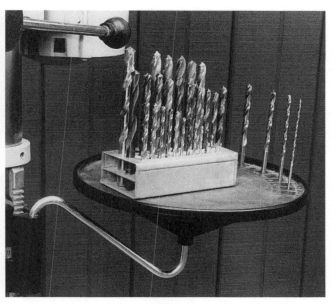

1-26 Wire wheels clean dirty or corroded metal parts. You can also use them to make wood look old and weathered. Hold the surface against the whirling wires, keeping the board moving. The wires will scrape away the soft springwood (the light-colored bands between the annual rings) much faster than the harder summerwood (the darker rings). This is just what happens when you leave wood outdoors — nature eats away the springwood faster than the summerwood. **Note:** Wear heavy work gloves when using wire wheels on the drill press. The wires will wear away the skin on your fingers even faster than springwood.

1-27 This plastic storage table fastens to the column of a drill press to hold bits and other accessories within reach. It also swings out of the way when you have large workpieces to drill. These are available from various sources, and you can make your own, following the plans in "Swing-Out Storage Tray" on page 21.

1-28 Several manufacturers offer auxiliary tables to extend the work surfaces of their drill presses. These are just sheets of particleboard or plywood that bolt to a drill press table. If you invest a little time, you can make a much better auxiliary table than you can buy. The shop-made table shown has mounting slots for a fence and a clamp, and an opening for drum sanders. It also tilts front to back and has a dust collection port. (See "Drill Press Table" on page 96 for plans and instructions.)

STORING THE CHUCK KEY

The most important drill press accessory is, of course, the chuck key. Consequently, according to Murphy's Law, it's also the one you are most likely to misplace. For nearly a century, craftsmen have experimented with ways to keep the key handy and have met with varying degrees of success. Here are two of the most noteworthy solutions.

1 **Probably the handiest place** to store the key is on the head of the drill press, where it's in plain view and within easy reach. You can use one of several devices to hang the key on the head — Velcro, hooks, holes, a shop-made bracket — but simplest is a strong magnet. Stick the magnet anywhere on the head that's easy to reach, then stick the chuck key to the magnet whenever you're not using it.

2 **Although the magnet is a** brilliant idea, it's only a partial solution for some woodworkers. Most chuck keys have legs, and even the strongest magnet won't keep them from wandering away from the drill press. When this is the case, you must tie your key to the drill press head. Once again, there are many ways to do this — wire, chain, shoelaces — but one of the cleverest is a retractable key chain. Attach the chuck key to the end of the key chain, then mount the reel on the drill press head with double-sided tape. The key chain becomes both a storage and a safety device: It ties the key to the press, and the spring-loaded reel prevents you from accidentally leaving the key in the chuck.

SWING-OUT STORAGE TRAY

Since you must constantly change bits and cutters when using the drill press, the handiest place to store them is out in the open. This simple storage table fastens right to the drill press column to keep the bits within easy reach but out of the way of drilling operations.

1 **This storage table is small,** but the "wedding cake" design holds hundreds of drill bits and cutters. Drill holes in the different tiers to hold the drilling tools that you use most often. Or, adapt the design of the tray to fit your drilling accessories. Mount the table on a swing-out arm. This arm pivots to allow you to place the table close to the work or out of the way, as needed. Additionally, the table rotates on its mount, helping you to find the tool you're looking for.

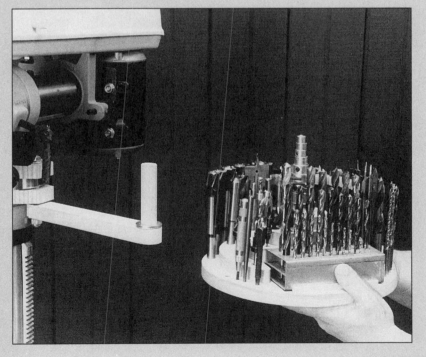

2 **When you need to remove the** table from the drill press, it simply lifts off of the arm. This makes it easy to get the table completely out of the way or to carry the bits and cutters to another location.

(continued) ▷

SWING-OUT STORAGE TRAY — CONTINUED

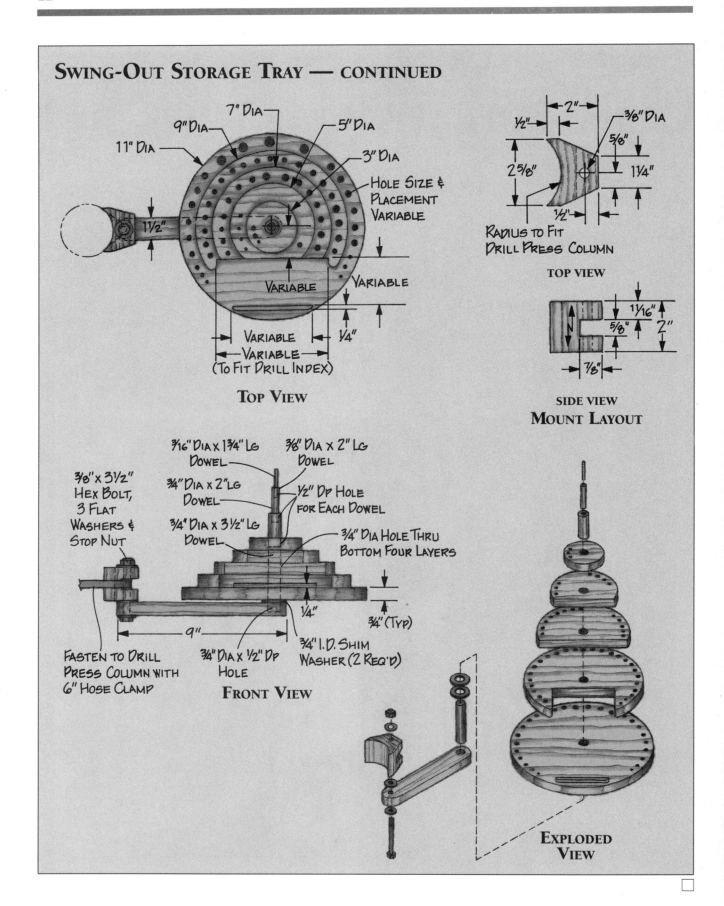

11" DIA
9" DIA
7" DIA
5" DIA
3" DIA

HOLE SIZE & PLACEMENT VARIABLE

1½"

VARIABLE
VARIABLE
¼"

VARIABLE
VARIABLE
(TO FIT DRILL INDEX)

TOP VIEW

2"
½"
⅜" DIA
5/8"
2 5/8"
1¼"
½"

RADIUS TO FIT DRILL PRESS COLUMN

TOP VIEW

1 1/16"
5/8"
2"
7/8"

SIDE VIEW
MOUNT LAYOUT

3/16" DIA x 1¾" LG DOWEL
⅜" DIA x 2" LG DOWEL
¾" DIA x 2" LG DOWEL
½" DP HOLE FOR EACH DOWEL
¾" DIA x 3½" LG DOWEL
¾" DIA HOLE THRU BOTTOM FOUR LAYERS

⅜" x 3½" HEX BOLT, 3 FLAT WASHERS & STOP NUT

¼"
¾" (TYP)

9"

FASTEN TO DRILL PRESS COLUMN WITH 6" HOSE CLAMP

¾" DIA x ½" DP HOLE

¾" I.D. SHIM WASHER (2 REQ'D)

FRONT VIEW

EXPLODED VIEW

2

DRILL PRESS KNOW-HOW

The drill press is one of the simplest and safest power tools to operate. The force of gravity and the pressure of the bit help to hold the work on the table, and the feed rate is limited by the cutting action of the tool. There is not much for you to do but keep your fingers at a safe distance, position the work under the bit, turn on the motor, and advance the quill.

To get the best possible results, however, you must properly align, adjust, and maintain your drill press. You should also know how to select bits and drill speeds for different materials, how to guide and secure the work, and how to keep the bit from chipping and tearing the wood. A little bit of drilling know-how will help you avoid some common pitfalls and make clean, precise holes.

ALIGNMENT AND ADJUSTMENT

RUN-OUT

To drill precise holes — holes that are accurately sized and positioned — your drill press must run straight and true. If the spindle is slightly bent or the chuck is mounted incorrectly, the drill bit may wobble, moving from side to side as it spins. The tip of the bit, instead of staying in the same place, will describe a circle. This condition is known as *run-out*.

Check the drill press for run-out when you first set it up, or whenever you experience problems with accuracy. Unplug the press and spin the chuck by hand, using a dial indicator or a feeler gauge to measure any side-to-side movement. (*SEE FIGURES 2-1 AND 2-2.*) For a drill press to provide the accuracy you typically need in woodworking, this movement should be less than .005 inch at the tip of the chuck.

2-1 To check drill press run-out with a dial indicator, unplug the machine and mount a large drill bit in the chuck. Position the indicator probe against the shank. Turn the front step pulley and spindle by hand. If the indicator needle moves more than .005 inch, the run-out is excessive.

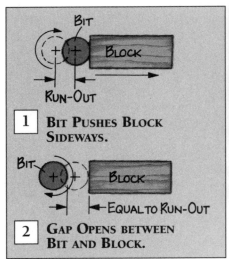

1 **BIT PUSHES BLOCK SIDEWAYS.**

2 **GAP OPENS BETWEEN BIT AND BLOCK.**

2-2 If you don't have a dial indicator, you can use a feeler gauge and a ½-inch-thick hardwood block. Once again, make sure the drill press is unplugged, and mount a drill bit with a ½-inch-diameter shank in the chuck. Raise the drill press table so most of the bit extends down through it. There should be about ¾ inch of the shank above the table. Rest the block on the table, against the drill shank. Rotate the step pulley and spindle by hand. As the chuck runs out, it will push the block sideways a short distance. As it runs back in, a space will open up between the drill press and the block. Hold the block on the table so it doesn't move, and use a feeler gauge to measure this space. If it's more than .005 inch, the run-out is excessive.

If there is excessive run-out, it may be caused by an improperly seated chuck. Try correcting the problem by tapping the "high side" of the drill bit shank (the side that runs *out*) with a hammer. *(See Figure 2-3.)* If that does not correct the problem, the run-out is probably due to a bent or worn part. If the drill press is new, ask the manufacturer to correct the problem. If it's an older tool, replace or rebuild the quill.

SPEED, TILT, HEIGHT, AND DEPTH

When you're satisfied the drill press is running as true as possible, you're ready to drill holes. There are four adjustments you must check whenever you work — drill speed, table tilt, table height, and drill depth.

DRILL SPEED

First of all, check that your drill press is set to the proper speed. *(See Figure 2-4.)* Different types of bits, bit diameters, and workpiece materials all require different drill speeds. Most woodworking operations are performed at speeds between 600 and 3,600 revolutions per minute (rpm); metalworking, between 100 and 900 rpm; and plastic, between 300 and 2,400 rpm. (Refer to the "Recommended Drill Press Speeds" chart on page 28.) For most operations, the rule of thumb is that the larger the drill bit and the harder the material, the slower you should run the drill press. There are some important exceptions to this, however. When working with wood, you must drill much more slowly parallel to the grain (into the end of a board) than perpendicular to it (into the board's face or side), or the bit will overheat. And with plastic, you should practice with scraps before drilling good stock. Plastic heats up quickly, retains the heat a long time, and expands ten times more than any other material. It also melts at a much lower temperature. For these reasons, it's trickier to drill than wood or metal.

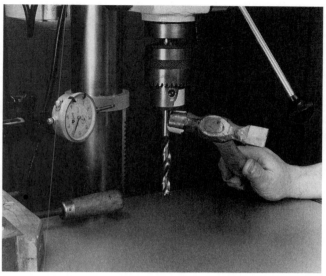

2-3 Excessive run-out may be caused by an improperly seated chuck or a sharp blow to the machine that knocked things out of whack. You can usually correct this problem with a few well-placed taps from a hammer. Once you've found the "high" side of the drill bit — the area on the circumference that pushes the dial indicator probe or the hardwood block farthest — mark it with a grease pencil or a bit of tape. Firmly *tap* (but don't whack) the high side of the bit shank with a hammer and check the run-out again. Repeat until the run-out is less than .005 inch. If after several taps the run-out is still outside the limits, you may have to repair or replace the quill. **Note:** Put a piece of masking tape over the face of the hammer to help protect the bit.

For Your Information

All materials have a *cutting speed* — a maximum rate at which the material can be machined without stressing or overheating the material or the cutter. There are hundreds of these speeds — one for every type of wood, plastic, metal, and metal alloy available — and they are usually given in feet per minute (fpm). These numbers are used to calculate the most efficient speeds for all sorts of power tools, including drill presses. If you take the time to look up a material's cutting speed in a machinist's reference, such as *Machinery's Handbook* (from Industrial Press), you can use this formula to calculate the maximum drill press speeds for that material:

$$N = (12 \times V) \div (3.1416 \times D)$$

Where:

V is the cutting speed in fpm

D is the diameter of the drill bit

N is the speed of the drill press in rpm

Drill speed, however, is not especially critical. You can drill a clean, accurate hole in a specific material with a specific bit at a wide range of speeds. The figures given in the chart are *recommended* speeds for drill presses with hand-operated quill feeds. If you drill a test hole and the edges look ragged, increase the speed for a smoother cut. If the bit overheats or the work burns, reduce the speed.

The best way to tell if your drill press is running at the proper speed is to observe the shavings as they come out of the hole. Wood shavings should be even in size, and they mustn't pack in the flutes or on the lifters. Metal and plastic shavings should appear as long, thin corkscrews or shavings similar to the wooden shavings produced by a sharp hand plane. (*SEE FIGURE 2-5.*)

2-4 To change the speeds of a drill press with a step-pulley speed changer, unplug the machine and change the positions of the V-belts on the steps. If the changer has just two pulleys, turn one pulley by hand and help the belt ride up or down the steps. If the changer has three pulleys, first release the belt tension, then reposition the belts. Remember to reapply the tension before turning on the machine.

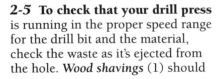

2-5 To check that your drill press is running in the proper speed range for the drill bit and the material, check the waste as it's ejected from the hole. *Wood shavings* (1) should be a uniform size and must not clog the bit. *Metal shavings* (2) and *plastic shavings* (3) should both be ejected in long, tightly coiled corkscrews. If the shavings are powdery or they become impacted around the bit, either the drill speed is incorrect or the bit is dull.

ACHIEVING ZERO RUN-OUT

Few tools are perfect; almost every drill press has a little run-out. By the same token, few drill bits are perfectly straight. They, too, run out just a little — especially bits with small diameters. With some patience, you can combine the chuck and the drill bit run-outs *to cancel each other out.*

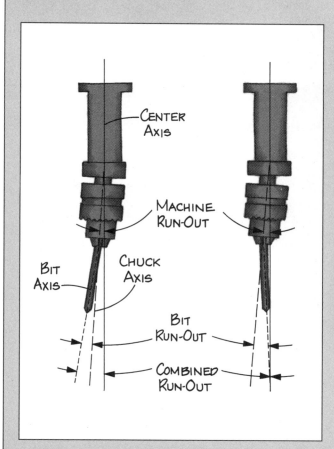

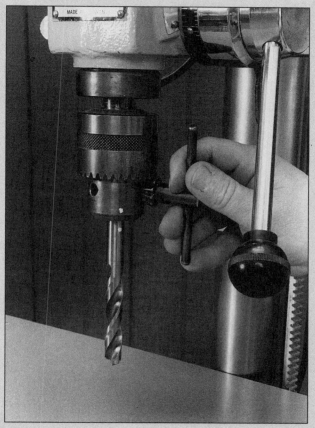

1 **With an indelible marker or** paint, mark one of the jaws in the chuck. Mount a drill bit, and turn on the drill press. Watch the bit's lead point — it probably will describe a small circle. Turn off the motor, loosen the chuck, and rotate the bit one-quarter turn in the chuck. Tighten the chuck and watch the tip of the bit again. Note whether the circle gets larger or smaller. Repeat, rotating the bit a fraction of a turn again. Continue to rotate in the same direction if the circle described by the point gets smaller; reverse direction if the circle grows larger.

2 **When you've positioned the** bit so the circle is as small as possible, make a mark on the bit's shank to correspond to the mark on the jaw. Thereafter, whenever you mount that bit, align the marks. This procedure takes a little extra time, but it ensures the smallest possible combined run-out and the highest degree of accuracy.

RECOMMENDED DRILL PRESS SPEEDS (IN RPM) FOR DRILL PRESSES WITH HAND-OPERATED QUILL FEEDS

This chart lists recommended speeds for drill presses with *hand-operated quill feeds*. They are lower than maximum cutting speeds because most craftsmen tend to feed the quill much more slowly than industrial presses with power feeds. To get a smoother cut and prevent the bit from overheating, the speeds have been reduced to compensate for a slow feed. Use these speeds as a starting point. Drill several test holes, then adjust the speeds up or down to compensate for your own drilling technique.

MATERIAL	DIAMETER	TWIST	BRAD-POINT	SPADE	BORING	FORSTNER	MULTI-SPUR
Softwoods	1/8"	4,800	3,600				
	1/4"	2,400	3,600	2,400		2,400	
	1/2"	1,200	2,400	2,400	2,400	2,400	1,200
	3/4"	1,200	1,800	1,800	2,400	1,800	1,200
	1"		1,200	1,800	1,800	1,200	1,200
	1 1/2"			1,200	1,200	900	1,200
	2"					600	900
	3"					300	600
Hardwoods	1/8"	2,400	3,600				
	1/4"	1,800	2,400	1,800		1,800	
	1/2"	1,200	1,800	1,800	1,800	1,800	1,200
	3/4"	700	1,200	1,200	1,800	1,200	1,200
	1"		900	1,200	1,200	900	1,200
	1 1/2"			900	900	600	900
	2"					300	600
	3"						300
Ferrous Metals	1/8"	900					
	1/4"	600					
	1/2"	300					
	3/4"	150					
	1"	100					
Nonferrous Metals	1/8"	2,400					
	1/4"	1,800					
	1/2"	1,200					
	3/4"	600					
	1"	300					
Plastics	1/8"	3,600	3,600				
	1/4"	2,400	2,400			1,200	
	1/2"	1,200	1,800			900	900
	3/4"	600	1,200			600	600
	1"		600			300	300
	1 1/2"						150
	2"						100

TABLE TILT

For most drilling tasks, the table must be perpendicular to the axis of the bit. You can check this quickly with a small engineer's square. (SEE FIGURE 2-6.) Or, if you require more accuracy, make a feeler gauge from a length of coat hanger wire or ⅛-inch-diameter metal rod. Make a sharp bend in the wire or the rod, and mount one end in the chuck. Turn the chuck by hand, using the other end to "feel" if the table is perpendicular to the axis of rotation. (SEE FIGURE 2-7.)

If you must feed the drill bit at an angle other than 90 degrees to the table, you have two choices — you can either build a jig to hold the work at the proper angle, or you can tilt the table. If you choose to tilt the table, gauge the angle with a protractor. (SEE FIGURE 2-8.)

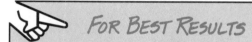

FOR BEST RESULTS

Make sure the table height lock is secure before squaring the table or adjusting the tilt.

2-6 To check that the table is perpendicular to the axis of the bit, use a small engineer's square and a spiral bit (twist bit or brad-point bit). With the square's base resting on the table, the arm should contact the bit all along its length. Check this angle at the front of the bit and along one side.

2-7 When accuracy is paramount, make a feeler gauge to check that the table is parallel to the bit's axis. Cut a 10-inch length of coat hanger wire or ⅛-inch-diameter metal rod, and make an 80-degree bend in it about 3 inches from one end. Unplug the drill press, mount the gauge's "short" end (nearest the bend) in the chuck, and adjust the table height so the other end just touches the table. Uncover the speed changer and turn the step pulleys and spindle by hand. Watch the feeler gauge as it rotates — the end should maintain contact with the table. If it breaks contact at any position, the table is tilted in that direction. Correct the table angle and check again.

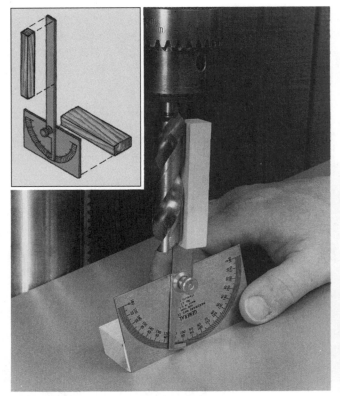

2-8 When you must tilt the table at an angle other than 90 degrees to the axis of the bit, use a protractor to gauge the angle. If you wish, modify the protractor as shown to make it easier to use with a drill press. Using double-faced carpet tape, attach a ½-inch-square strip to the *front* of the arm and a ½-inch-thick, 1-inch-wide strip to the *back* of the head. Make sure the edges of the strips are flush with the edges of the parts you stick them to.

TABLE HEIGHT

You must allow adequate space between the table and the tip of the bit for the work, but not so much space that it requires most of the stroke just to get the bit to the work. To adjust this space, raise or lower the table. Release the table lock and turn the table height adjustment crank. (*SEE FIGURE 2-9.*) When the table is at the proper level, secure the lock again.

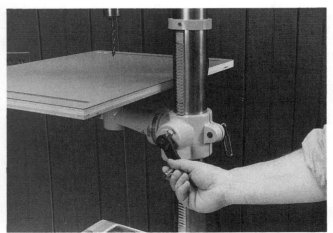

2-9 Most newer, high-quality drill presses use a rack-and-pinion table height adjustor. The crank turns the pinion gear, which climbs up or down the rack on the column, raising or lowering the table. The rack is not fastened to the column; it's held loosely at the top and bottom in grooved collars. This lets you swing the table around the column as well as change its height. When you've positioned the table where you want it, lock it down on the column.

TRY THIS TRICK

To quickly raise and lower the work surface, make two or three ¾-inch plywood spacers, roughly the same size as the table or a little smaller (to make room for a fence). Instead of using the rack and pinion to raise or lower the table 1 or 2 inches, add or subtract spacers. Since most craftsmen do the bulk of their drilling in a 2-inch height range, you'll rarely have to change the level of the table.

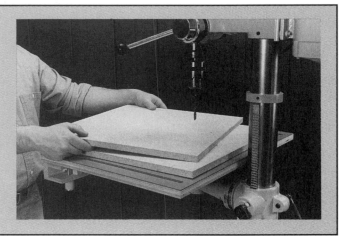

TILTING WORK SURFACE

The tilting mechanisms on most drill press tables are notoriously inconvenient. The table usually pivots on a large bolt that's tucked away under the table arm, where it's difficult to reach with a wrench. Additionally, the table tilts from side to side, making it difficult to hold long boards and round stock. It's a simple matter to build a tilting fixture that's more capable and more convenient to use.

1 **This fixture uses a unique** system of *tamboured wedges* that makes it possible to adjust the work surface to a precise angle instantly, without measuring. Each wedge is cut at a 7½-degree angle, and there are eight wedges to a side, letting you set the work surface at 7½, 15, 22½, 30, 37½, 45, 52½, and 60 degrees. Simply fold the wedges you don't need out of the way, and rest the remaining wedges over the dowel pins on the sides of the fixture.

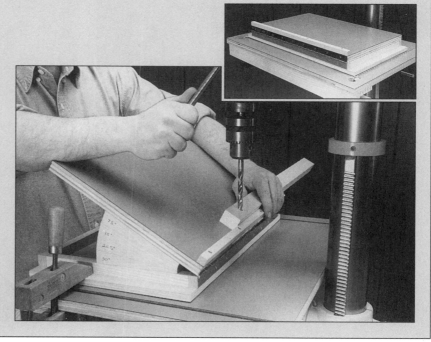

2 **Clamp the tilting work sur-**face to the drill press table so it holds the work in the proper position under the bit. If necessary, insert a scrap between the work and the work surface to keep from drilling through the fixture. When you finish the operation, fold the tilting work surfaces flat for storage.

(continued) ▷

TILTING WORK SURFACE — CONTINUED

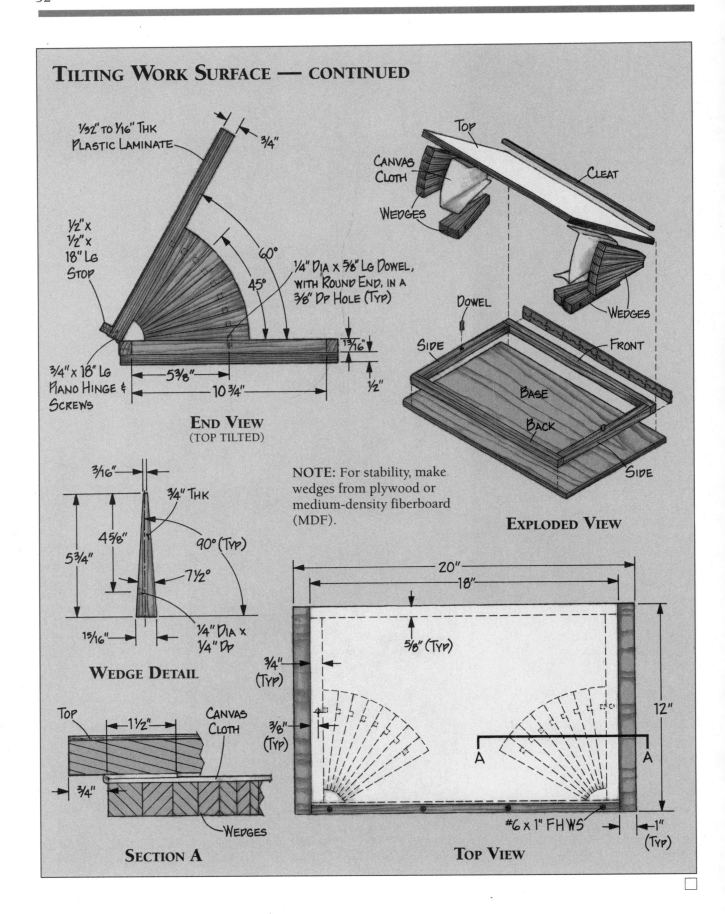

NOTE: For stability, make wedges from plywood or medium-density fiberboard (MDF).

END VIEW
(TOP TILTED)

WEDGE DETAIL

SECTION A

EXPLODED VIEW

TOP VIEW

DEPTH STOP

On most drilling operations, you have to limit the feed. You must either stop the drill bit before it exits the workpiece or prevent it from boring through the backup scrap and into the table. The depth stop performs both of these functions. (*See Figures 2-10 and 2-11.*) You can also use the depth stop to prevent the quill from retracting into the head. (*See Figure 2-12.*)

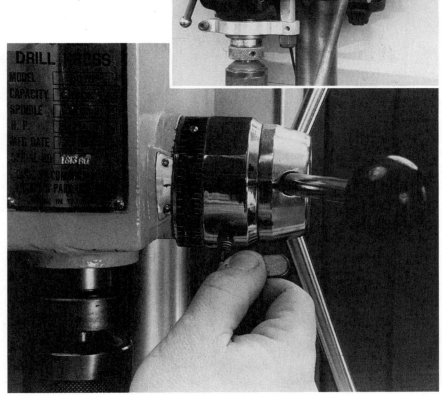

2-10 **To drill a hole of a specific** depth, extend the quill until the tip of the bit touches the workpiece — this is the "0" position. If you have a newer drill press with a rotating stop, loosen the lock on the depth stop, turn the scale until the desired measurement aligns with the indicator, and secure the lock. If you have an older model, turn the nuts on the threaded post until they rest against the tabs on the head casting. **Note:** The scales on most depth stops are not especially accurate. The depth stop's accuracy may also be adversely affected by the size of the lead point on the bit. If you need to drill a hole to a *precise* depth, make a test hole in a scrap that's the same thickness as the workpiece. Measure the depth of the hole, adjust the depth stop as needed, and make another test hole. Don't drill the workpiece until you're sure the setup is accurate.

2-11 **You can also use the depth** stop to prevent the drill bit from boring into the table when drilling through a workpiece. Place a scrap of wood or plywood on the table to back up your work. Extend the quill until the tip of the bit touches the scrap, then set the depth stop to drill partway into the scrap. Position the workpiece on top of the scrap and drill the hole.

2-12 You can use the depth stop to keep the quill from retracting into the head as well. Extend the quill to the position you want, rotate the depth stop dial *clockwise* until it stops, and secure the lock. **Note:** While this technique holds the vertical position of the quill, it does *not* prevent it from moving completely. There is still too much horizontal and vertical play to use the drill press for anything but extremely light-duty operations. If you use it for heavy work, like routing or sanding, you risk damaging the machine — as the quill moves back and forth, it will enlarge its channel in the head casting. See "Quill Locks and Side Thrust" on page 93.

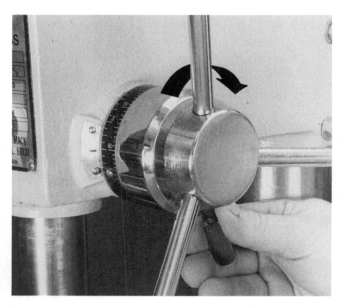

DRILLING A HOLE

PREPARING TO DRILL

Once you've made all the necessary adjustments, there are still a few things to do before you can drill.

Mounting the Drill Bit — Select the drill bit that is best suited for the material and the operation. While twist bits and brad-point bits will perform most drilling tasks, other types may produce better results. When the choice is not clear, drill several test holes with different bits.

When mounting the bit in the chuck, it's good shop practice to tighten the jaws *twice*. There are three places on the chuck where you can insert the nose of the chuck key to turn the ring gear. Tighten the jaws at two different locations — this helps prevent the bit from slipping.

Laying Out the Work — Measure the horizontal and vertical position of each hole you want to drill, and draw two intersecting lines on the workpiece to indicate its position. Where the lines cross, create a small hole or indentation with an awl or a punch. (*SEE FIGURE 2-13.*)

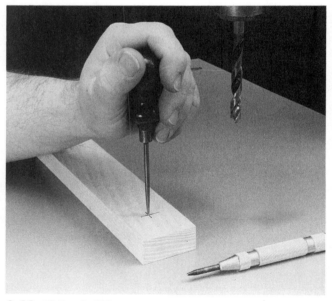

2-13 If the drill bit you're using has a long lead point, create a small indentation at each hole location with an *awl* or a *prick punch*. This step isn't absolutely necessary, but the indentations help center the point and start the bit. When using a twist bit, make the indentation with a *center punch*. This leaves a broader indentation than a prick punch, better fitted to the twist bit's broad tip. The indentation prevents the hard summerwood (the dark rings in the board) from deflecting the bit and causing it to wander off the mark before it begins to bite into the wood.

A SAFETY REMINDER

Never leave the chuck key in the chuck — the drill press will fling it out (possibly at you) if you accidentally turn the machine on. Get in the habit of returning the key to a handy (but safe) location when you're not using it.

Positioning and Securing the Work — Place the work under the bit, advance the quill so the end of the bit almost touches the work, and align the indentation that marks the hole with the tip of the bit. Whenever possible, use a fence to help hold the work in position. (*SEE FIGURE 2-14.*) In many cases, you will want to clamp the workpiece to the table or the fence. (*SEE FIGURE 2-15.*) When drilling small parts, large holes, or materials that tend to catch on the bit, it's much safer to secure the work to the drill press. Securing the work also ensures accuracy.

A SAFETY REMINDER

One of the greatest dangers in using a drill press is that a workpiece might catch on the bit and start spinning. If this happens, a small board can build up enough momentum to break your wrist. A piece of sheet metal becomes a whirling blade that will sever a finger. Your best defense is to always make sure that the workpiece is properly braced or clamped before you drill. But if the bit should catch unexpectedly and the workpiece is yanked from your hands, do *not* try to grab it. Get your hands clear immediately and turn off the machine.

2-14 **Whenever you can, clamp a** fence to the drill press table and use it to help hold the work. It's much easier to keep the work in one position when you can press it against the table *and* the fence than it is to hold the work steady on the table only. This arrangement also prevents the stock from spinning, should it catch on the bit.

2-15 **It's not always enough to use** the table and the fence to help steady the work. Oftentimes, you must fasten the work to the table, the fence, or another piece of wood before you can drill it safely. For example, when drilling a small workpiece, fasten it to a larger scrap with double-faced carpet tape and hold the scrap against the fence (1). When drilling a large hole or working with a material that tends to catch on the bit (such as thin metal), clamp the workpiece to the table (2). When working with a round or curved workpiece, make a jig to hold it and fasten both the jig and the workpiece to the table or the fence (3).

BASIC DRILLING TECHNIQUE

When the work is secure, you're ready to drill.

Starting the Hole — Turn on the drill press and *slowly* advance the bit until the lead point touches the wood. The point should touch the indentation you've made. If it doesn't, retract the drill and realign the workpiece. When you're satisfied the workpiece and the bit are aligned accurately, *slowly* advance the drill until the lips begin to cut, then gradually increase the feed pressure and complete the hole.

Some craftsmen advance the bit so the lead point is resting in the indentation before they turn on the machine. As long as both the chuck and the bit are running reasonably true, this ensures the hole is accurately positioned.

Whatever method you use, it's important to advance the bit very slowly until the hole is started. If you increase the feed rate too quickly, the bit may be deflected to one side.

Through Holes — When drilling through the stock, back up the workpiece with a scrap of wood. There are two reasons for this. First, the scrap protects the drill press table and keeps the bit from biting into it. Second, it supports the wood fibers as the bit exits the stock and helps prevent them from tearing or chipping. (*See Figure 2-16.*)

A BIT OF ADVICE

A *foot switch* adds safety, versatility, and convenience to your drill press operations by enabling you to keep one hand on the work and the other on the controls at all times. You can start the drill press while positioning a workpiece and operating the quill feed. And should something go wrong, all you have to do is step away from the machine to shut it down.

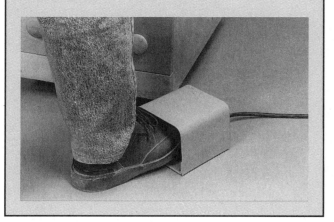

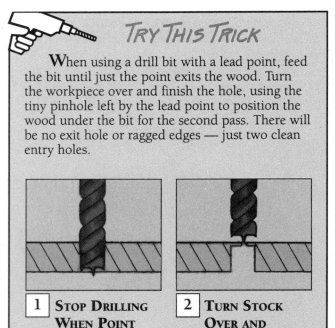

TRY THIS TRICK

When using a drill bit with a lead point, feed the bit until just the point exits the wood. Turn the workpiece over and finish the hole, using the tiny pinhole left by the lead point to position the wood under the bit for the second pass. There will be no exit hole or ragged edges — just two clean entry holes.

1 STOP DRILLING WHEN POINT EXITS.

2 TURN STOCK OVER AND FINISH HOLE.

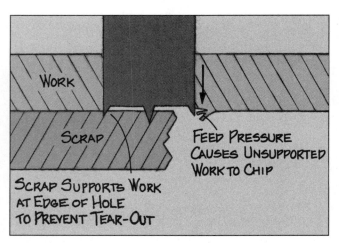

WORK

SCRAP

FEED PRESSURE CAUSES UNSUPPORTED WORK TO CHIP

SCRAP SUPPORTS WORK AT EDGE OF HOLE TO PREVENT TEAR-OUT

2-16 The edges of entrance holes (where a bit enters the wood) are usually clean and crisp, but the wood fibers around exit holes tend to tear out, making the holes look ragged. To support the fibers and prevent tear-out, back up the workpiece with a scrap. **Note:** You must move this scrap around as you drill each hole so there's always a fresh, flat surface where the bit exits the stock. Otherwise, the wood fibers won't be properly supported. Eventually, you must replace the scrap with a fresh backup surface.

TRY THIS TRICK

Masking tape "flags" make excellent depth indicators, particularly for small drill bits. Wrap the masking tape around the bit to create a flag where

you want it to stop. Then drill the hole, watching the sawdust as it's ejected. When the flag begins to wipe away the dust, stop drilling.

Stopped Holes — When drilling a stopped hole, you don't have to worry about backing up the workpiece, since there is no exit hole. But you must set the depth stop, as described in "Alignment and Adjustment" on page 24, to limit the stroke. Or, fasten a stop collar to the bit to keep it from penetrating completely.

AFTER MAKING THE HOLE

When the hole is completed, there's one more thing to do.

Cleaning Up — Carefully brush away the wood shavings before starting the next hole or the next operation. If this waste is allowed to accumulate, it may affect the accuracy of the setup, preventing the work from butting up against the fence or sitting flush on the table. If you're drilling through holes, sawdust can also cause tear-out by holding the workpiece a short distance above the backup board and keeping the board from properly supporting the wood fibers around the exit hole.

DRILL PRESS MAINTENANCE

CARING FOR THE DRILL PRESS

To maintain a drill press in good working order, brush or blow away the sawdust from the table and the head after every use. Be especially careful to keep the quill and the chuck clean. Fine sawdust will ruin the bearings, prevent the chuck from operating smoothly, and cause the bits to slip.

If the jaws of the chuck seem to stick and will not grip the bits securely, there is probably dirt or dust in the chuck. To clean the chuck thoroughly, remove it from the quill by breaking the friction lock between the tapered surfaces of the chuck and the spindle. (*SEE FIGURES 2-17 AND 2-18.*) Retract the jaws into the chuck, and clean out any grime with a toothbrush and mineral spirits. Move the jaws in and out several times to loosen more grime, and continue to clean. (*SEE FIGURE 2-19.*) When the jaws move freely and the chuck is as clean as you can make it, inspect the clamping surfaces of the jaws. If they are scarred or galled, *lightly* file them with a needle file. (*SEE FIGURE 2-20.*) Apply a few drops of light machine oil to the top and bottom of the sleeve, and replace the chuck on the drill press. (*SEE FIGURES 2-21 AND 2-22.*)

There is some controversy about whether or not you should oil the chuck. According to some, the spinning chuck will fling the oil out at you and the workpiece. Oil also attracts dust, and the two mix to become grime. This, in turn, prevents the parts from moving smoothly. While these concerns are valid, both problems are the result of *too much* lubrication. If you apply the oil *sparingly,* there won't be any excess to be flung out or to turn to grime. But if you'd rather not oil the chuck, you can use silicone, graphite, and other dry lubricants.

You should, however, refrain from lubricating the *inside* of the chuck. Oil or any other lubricant may prevent the jaws from gripping the bits securely. It will also prevent the tapered surfaces from locking when you mount the chuck back on the quill. What about the mineral spirits used for cleaning — won't they have the same effect? Mineral spirits do act like lubricant while they last, but they quickly evaporate, leaving bare metal surfaces.

Every few months, apply a little paste wax to the table, quill, and column, then buff it out thoroughly. This prevents rust and helps the parts move smoothly. When buffed, the wax provides lubrication, making it easier to slide the table up and down the column or the quill in and out of the head.

 ## A BIT OF ADVICE

Although it's possible to take apart some chucks by pressing the sleeve off the chuck body, it's not necessary to do so just to clean them. Furthermore, many newer chucks are permanently assembled. If you try to press them apart, you may ruin them.

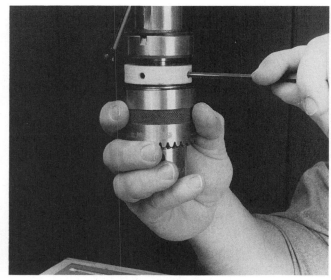

2-18 Some tool manufacturers in-stall a threaded collar directly above the chuck to help remove it. Insert the key in the chuck and grip it with one hand to keep the chuck from turning. With the other hand, use a wrench to turn the collar counter-clockwise. The collar will push the chuck off the spindle. **Note:** If the chuck is stubborn, tap it with a hammer while applying pressure with the threaded collar or wedges. This will help to loosen the friction lock.

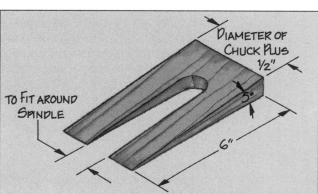

DIAMETER OF CHUCK PLUS ½"

TO FIT AROUND SPINDLE

5°

6"

2-17 Most chucks are mounted to the drill press spindle by a *locking taper.* The end of the spindle is ground to a standard-size taper, and the chuck body is machined to match it. When the spindle is inserted in the chuck with a little pressure, the tapered surfaces lock together. To remove the chuck, you must break this friction lock. If there is no collar directly above the chuck, make a pair of hardwood wedges and press them into the space between the chuck and the spindle with a clamp. As the wedges slide together, they will push the chuck off the spindle.

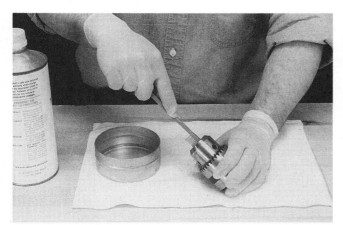

2-19 Once the chuck is free, turn the ring gear until the jaws retract into the chuck as far as they will go. Scrub out the inside of the chuck with a toothbrush and mineral spirits. While the chuck is still wet with spirits, turn the ring gear back and forth several times, advancing the jaws until they come together and then retracting them completely. This will loosen more grime. Scrub the inside of the chuck again and repeat until it's as clean as you can make it.

2-20 When the chuck is clean, inspect the gripping surfaces of the jaws. If there are any nicks or burrs, file them smooth with a needle file. Use a light touch and just eliminate the high spots; if you remove too much metal, the jaws won't be even and they will hold the bit off-center.

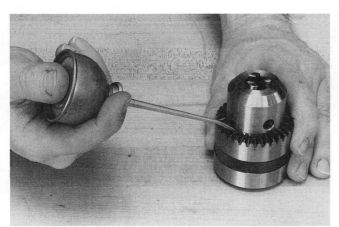

2-21 Apply one or two drops — *no more* — of light machine oil (10W) to the seam at the top and bottom of the sleeve. Spin the ring gear back and forth several times to spread the oil over the moving parts. Be careful not to get any oil on the inside of the chuck. If the jaws are lubricated, they may not grip the bits properly. And if the tapered hole is lubricated, the chuck may not lock onto the spindle.

APPLY LUBRICANT HERE...

AND HERE

2-22 When the chuck is clean and properly lubricated, insert the tapered end of the spindle into the tapered hole in the chuck. Place a scrap board on the drill table and extend the quill until the nose of the chuck rests against the board. Press down firmly on the quill feed and turn the machine on for several seconds. This will lock the tapered surfaces together again.

DRILL PRESS FOOT SWITCH/QUILL PEDAL

More often than not, drill press operations seem to require three hands. It's impossible to hold the work, advance the quill, and turn the machine on all at the same time. This foot switch/quill pedal frees *two* hands, letting you operate the on/off switch *and* the quill with a single foot. If necessary, you can keep *both* hands on the work.

Both the foot switch and the quill pedal are mounted to a small box that can be positioned almost anywhere near the drill press. (This box is weighted with lead shot so it stays put.) The quill pedal is attached to the quill feed by a cable and a pulley, and the switch is wired to an electrical outlet on the back of the box.

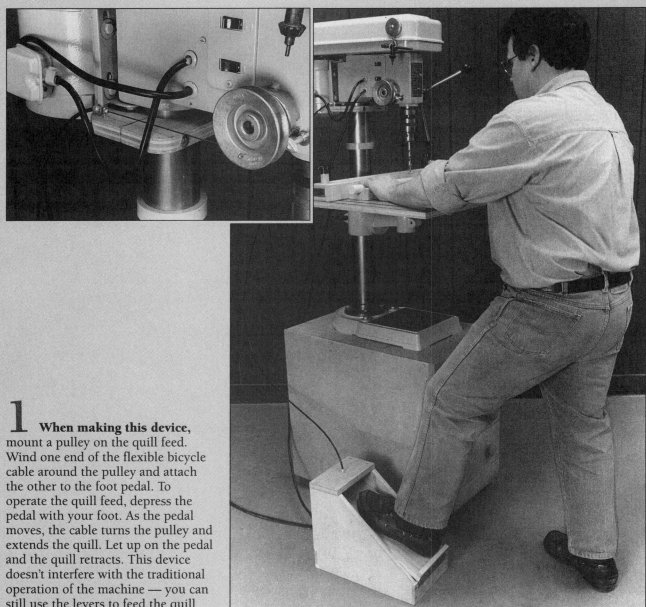

1 **When making this device,** mount a pulley on the quill feed. Wind one end of the flexible bicycle cable around the pulley and attach the other to the foot pedal. To operate the quill feed, depress the pedal with your foot. As the pedal moves, the cable turns the pulley and extends the quill. Let up on the pedal and the quill retracts. This device doesn't interfere with the traditional operation of the machine — you can still use the levers to feed the quill when necessary.

2 **Make the pedal 2 inches** wider than your foot, and mount a "Momentary On" switch 1 inch away from one edge of the quill pedal. (If you plan to operate the pedal with your right foot, position the switch closest to the right edge. If you will use your left foot, move the switch to the left.) Connect the switch to control the power to an electrical outlet at the back of the box. By plugging the drill press into this outlet and throwing the machine's switch to "On," you can use the pedal switch to turn the drill press on and off. When your foot is over the switch, light pressure will start the drill press and heavier pressure will extend the quill while the drill press is running. To turn the drill press off, take your foot off the pedal or throw the machine's switch to "Off." Move your foot to one side so it doesn't cover the switch, and you can feed the quill without turning on the machine.

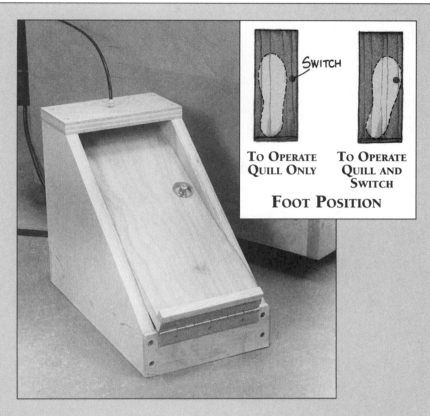

To Operate Quill Only

To Operate Quill and Switch

FOOT POSITION

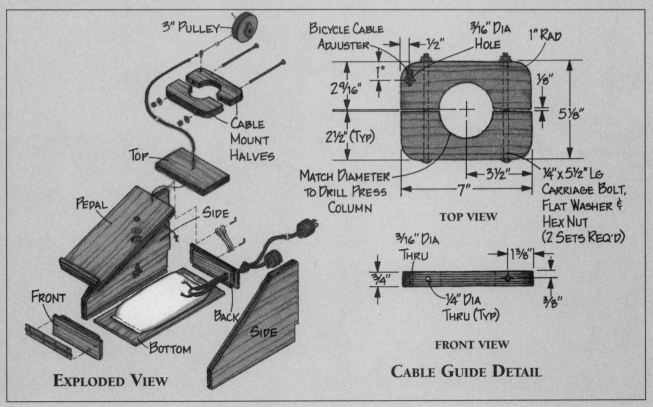

EXPLODED VIEW

CABLE GUIDE DETAIL

TOP VIEW

FRONT VIEW

(continued) ▷

DRILL PRESS FOOT SWITCH/QUILL PEDAL — CONTINUED

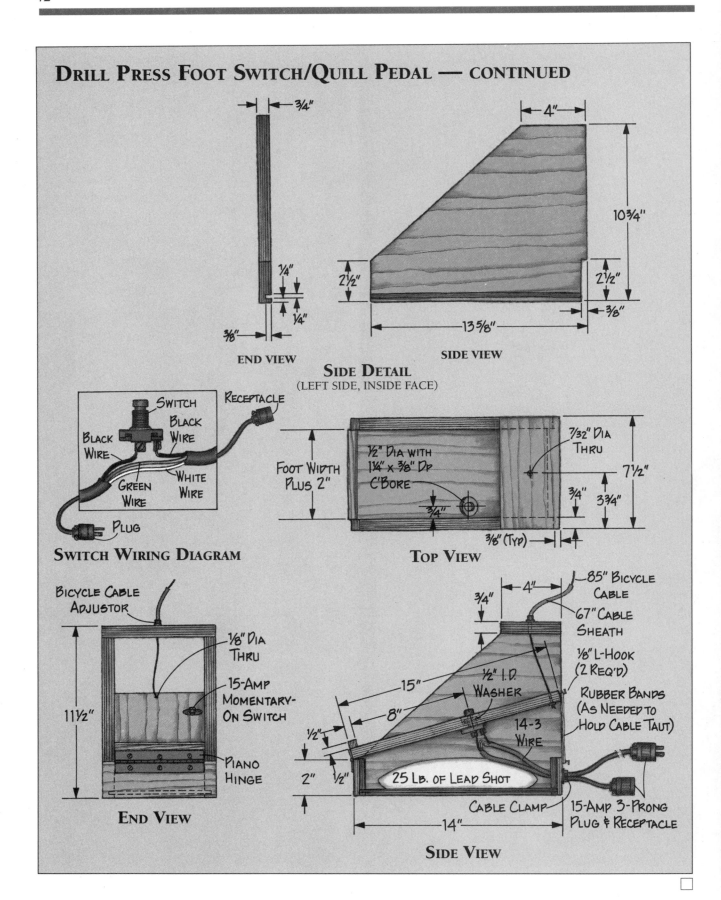

END VIEW

SIDE VIEW

SIDE DETAIL
(LEFT SIDE, INSIDE FACE)

SWITCH WIRING DIAGRAM

TOP VIEW

END VIEW

SIDE VIEW

CARING FOR DRILL BITS

When storing drill bits, separate them in a divided box or a drill index and organize them by size. *(SEE FIGURE 2-23.)* Don't throw them together in a drawer where they can rattle around and bang against one another — the cutting edges will become nicked and gouged. If you live in a humid climate, keep a cake of camphor in the cabinet or tool box with the bits. Camphor slowly evaporates and condenses on nearby metal surfaces, coating them with a thin, oily film that prevents rust.

When a bit becomes dull, it's best to have it sharpened by a professional sharpener, particularly brad-point bits, Forstner bits, and multi-spur bits. The cut-ting surfaces on these drill bits are ground at complex *symmetrical* angles, and this symmetry must be maintained when they are sharpened or else the bits will no longer cut accurately. Most small woodworking shops don't have the sharpening equipment necessary to ensure symmetry. However, you can lightly hone or "touch up" your drill bits when they start to grow dull. As long as you don't remove too much metal, this will restore the edge without affecting the symmetry, and it extends the time between professional sharpenings. Refer to "Sharpening Drill Bits" on page 44 for sharpening instructions on specific types of bits.

2-23 When storing drill bits, separate them to protect the cutting edges and organize them so you can easily find the size you need. If you must store the bits lying down, a *divided box* (left) works well. If you have room to stand them up, use a *drill index* (right). There are many commercial indexes available, and you can make your own.

FOR BEST RESULTS

Purchase an *auger bit file* (1) and a *slip stone* (2) to touch up drill bits. These two sharpening tools will restore the cutting edges of every common type of bit.

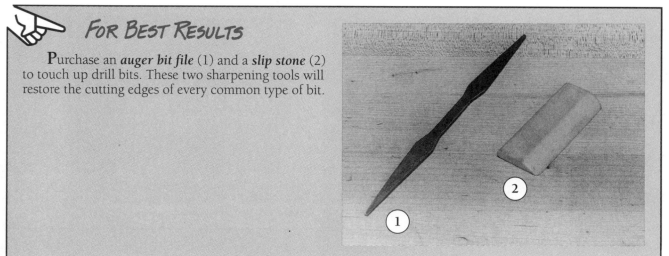

SHARPENING DRILL BITS

Although the procedure for sharpening each type of drill bit differs, there are two important rules to keep in mind no matter what kind of bit you're working on. First, count your strokes and file or hone each side of the bit evenly to keep the cutting edges as symmetrical as possible. And second, *never* sharpen the circumference (outside surface) of the bit — you'll change the diameter.

TWIST BITS

The lips of twist bits are ground at compound angles. As they come from the manufacturer, the lips are 118 degrees from one another and come together in a point. The areas behind the cutting edges — the *heels* — are angled back from the lips at 12 degrees to clear the wood's surface. (This is referred to as the *clearance angle*.) You can buy commercial sharpening fixtures that will hold the bit in the proper position to hone these angles, or you can make your own.

To use the shop-made jig shown, clamp the bit between the two halves so the lips are parallel to the inside edges. Adjust the bit's vertical position so the heels are flush with or just a hair above the angled surfaces at the top of the jig. Lightly hone the heels, using the jig's surfaces as a guide.

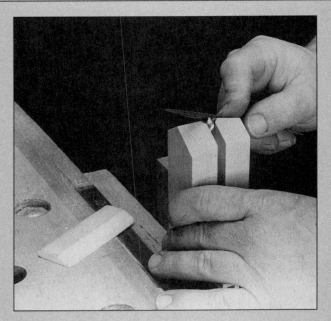

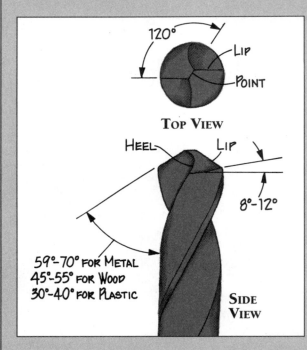

120°

LIP

POINT

TOP VIEW

HEEL   LIP

8°–12°

59°–70° FOR METAL
45°–55° FOR WOOD
30°–40° FOR PLASTIC

SIDE VIEW

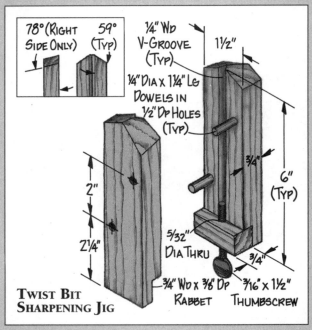

78° (RIGHT SIDE ONLY)   59° (TYP)

¼" WD V-GROOVE (TYP)

1½"

¼" DIA X 1¼" LG DOWELS IN ½" DP HOLES (TYP)

2"

2¼"

¾"

6" (TYP)

5/32" DIA THRU

¾" WD X ⅜" DP RABBET

¾"

3/16" X 1½" THUMBSCREW

TWIST BIT SHARPENING JIG

BRAD-POINT BITS

The lips of a brad-point bit are perpendicular to the axis, and the areas behind the lips — the *lands* — are usually angled at about 70 degrees. Before you touch up a brad-point bit, measure the land angle with a protractor. Make the jig shown, beveling the top surfaces to match the slope of the lands.

Clamp the bit in the jig, aligning the lands with the beveled surfaces. Take a good look at your auger file — note that one end has safe edges while the other end has safe faces. (A *safe* file surface has no teeth.) Use the file end with safe faces to touch up the lands, resting the file on the beveled surfaces of the jig to guide it. Then touch up the *inside* surfaces of the spurs, using the file end with the safe edges.

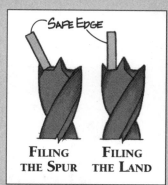

FILING THE SPUR   FILING THE LAND

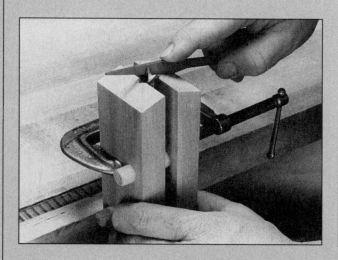

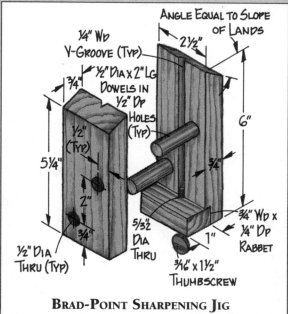

BRAD-POINT SHARPENING JIG

SPADE BITS

The lips of a spade bit are ground at 10 degrees. To sharpen it, make the jig shown and clamp the bit in it. Adjust the bit's vertical position so the lips are just a hair above the top surface of the jig. Using these surfaces as a guide, file each lip evenly. After filing the lips, hone the sides of the lead point with a slip stone. Hold the stone so it's angled back about 10 degrees to match the clearance angle of the point.

(continued on next page)

(continued) ▷

SHARPENING DRILL BITS — CONTINUED

SPADE BITS — CONTINUED

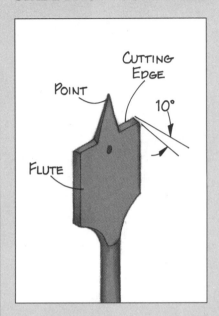

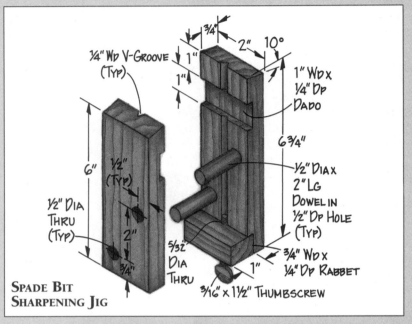

SPADE BIT
SHARPENING JIG

BORING BITS

The lifting faces of a boring bit are usually ground 30 degrees from the axis, and the clearance angle of the lips is 10 degrees. To touch up this bit, sharpen the lifting faces with a slip stone, pressing the stone flat against the surface. Lightly hone the lips, angling the stone to match the clearance angle. Touch up the inside surface of the spurs with an auger file in the same manner as the spurs on a brad-point bit.

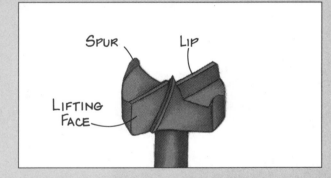

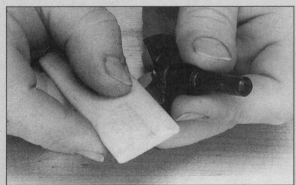

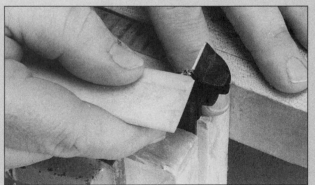

FORSTNER BITS

The lifting faces of a Forstner bit are typically ground at 45 degrees to the axis, and the clearance angle behind the lips is 10 degrees. Additionally, the cutting rims are sharpened at about 45 degrees on the inside surfaces. Hone the lifting faces with a slip stone, keeping it pressed flat against the surfaces. Then hone the lips, using the clearance notches on each side of the bit to reach them. Finally, use the rounded edge of the slip stone to sharpen the inside edge of the rim. Roll the bit back and forth in your fingers while holding the rim against the stone.

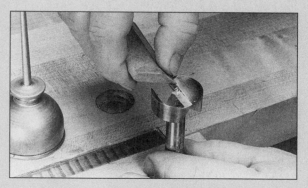

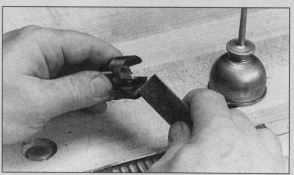

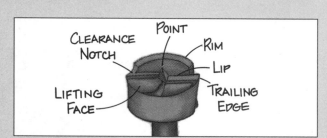

MULTI-SPUR BITS

The lifting face and the clearance angle behind the lip of a multi-spur bit are ground in much the same manner as a Forstner bit. The rim, however, is serrated with several spurs or sawteeth. Sharpen the lifting face first, holding a slip stone flat against it. Next, touch up the lip, using the clearance notch on the rim to reach it with the stone. Sharpen the leading face of each spur with an auger file, holding the file at a slight angle to match that of the teeth.

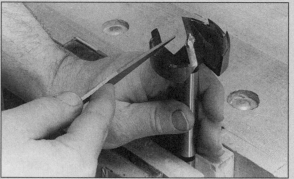

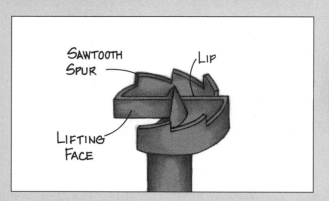

3

DRILLING AND BORING

In the course of a typical woodworking project, you must bore
many different types and sizes of
holes — deep holes, angled holes,
holes to fit screws and other hardware. Some of these holes require
special cutters; others can be
drilled with general-purpose bits.
You may have to drill holes across
the wood grain or parallel to it, or
bore through altogether different
materials, such as metal and plastic.
While all drilling operations are
similar, every type of hole, bit, and
material requires a slightly different
technique.

DRILLING CROSS GRAIN

When working with wood, most holes are drilled across the grain. Ordinarily, these are relatively small, perpendicular to the surface, no deeper than the stroke of your drill press — and easy to make. However, now and then you must drill a hole that's out of the ordinary.

DEEP HOLES

When drilling holes that are deeper than the stroke of your drill press, there are two techniques you can use.

If the hole is deeper than the stroke but shorter than the length of the bit, first drill the hole as deep as possible in a single pass, then *raise the table* (or the work) and make another pass. Repeat until you have made the hole as deep as needed. (*SEE FIGURES 3-1 AND 3-2.*)

A SAFETY REMINDER

Turn off the drill press and let it come to a complete stop before raising the table and the work.

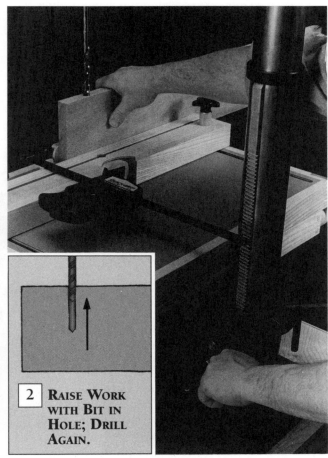

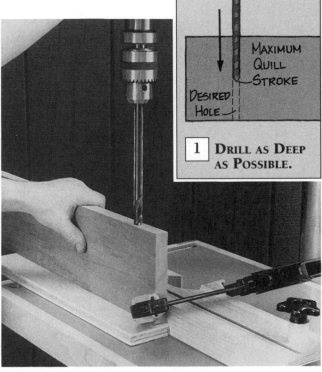

1 DRILL AS DEEP AS POSSIBLE.

2 RAISE WORK WITH BIT IN HOLE; DRILL AGAIN.

3-1 To drill a deep hole, use a bit that's longer than the desired hole depth, if one is available. If the bit is longer than the stroke of the drill press, make the hole in several passes. To begin the hole, position the workpiece under the bit and drill as deep as the machine will let you. Retract the bit and turn off the drill press.

3-2 When the bit has come to a complete halt, raise the table and the workpiece until the tip of the bit almost touches the bottom of the hole. Lock the table in place, turn on the drill press, and make another pass. Repeat this procedure until you have drilled through the workpiece or made the hole as deep as you require. **Note:** When you turn on the drill press with the bit inside the hole, there is a chance the bit will catch on the stock. To prevent this, secure the workpiece.

To drill a hole deeper than the length of the drill bit, drill two holes that meet in the center of the stock. After drilling the first hole, align a pin with the bit and secure it to the table. Use the pin to position the work under the bit to drill the second hole. (*SEE FIGURES 3-3 AND 3-4.*)

LARGE HOLES

There are three different cutters or bits that you can use to make large holes (over 1½ inches in diameter). The choice depends on the results you're after and the materials you're working with. If you want a smooth-sided hole, use a *multi-spur bit*. If you need to drill a hole quickly, or if you're working with construction-grade materials, use a *hole saw*. (*SEE FIGURES 3-5 AND 3-6.*) Both of these accessories come in ⅛-inch increments up to 3¼ inches, and ¼-inch after that. To drill odd-size holes, use a *fly cutter*. (*SEE FIGURE 3-7.*)

> ## TRY THIS TRICK
>
> **W**hen using a hole saw, many craftsmen prefer to cut most of the way through the stock, stopping when the pilot bit is through the wood. Then they turn the stock over to finish the hole, using the pilot hole to center the saw for the second cut. This prevents the tear-out that occurs when the saw exits the wood and makes it much easier to remove the waste plug from the saw.

When making large holes, always clamp the work to the drill press table or fence; don't try to hold it. The larger the bit, the greater the chance that it will catch the workpiece and spin it. If you are drilling a large hole in a small piece and there isn't room to clamp it, stick the work to a large scrap with double-faced carpet tape; then clamp the scrap to the drill press.

3-3 If you must drill a hole through stock that's thicker than the bit is long, make *two* holes that meet in the middle of the stock. Drill the first hole, and turn the stock over to make the second. To position the work so the holes meet, make a simple alignment jig. In a large, flat board, mount a dowel the same diameter as the holes you're drilling. To align the dowel to the bit, drill a hole through a smaller block, the same diameter as the dowel. Slip the block onto the dowel. Extend the quill and adjust the dowel's position until the bit can be buried in the block, too. Clamp the jig to the table.

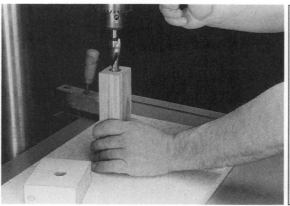

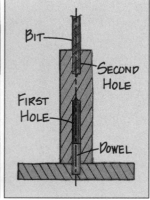

3-4 Rest the workpiece on the scrap, placing the first hole over the dowel. As long as the table is square to the bit and the dowel is square to the table, this will center the first hole directly under the bit. When you drill the second hole, it will meet up with the first.

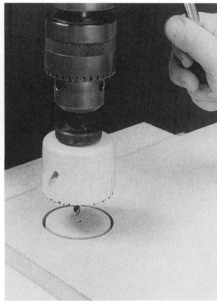

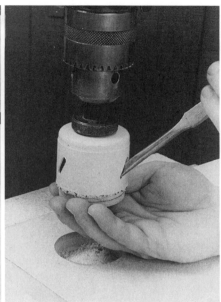

3-5 When you must drill precise holes over 1½ inches in diameter, the best cutter for the job is often a multi-spur bit. It leaves a smooth-sided hole but doesn't have as much tendency to "grab" the stock as a large-diameter Forstner bit. Adjust the drill press to run at a low speed, and feed the bit very slowly.

3-6 If you need to drill a large hole quickly, and aren't especially concerned about its precision or the condition of the edges, use a hole saw. As with a multi-spur bit, run the drill press at a low speed and feed the saw slowly. These tools have no way to eject the waste, so you must retract the saw often and clear away the saw-dust to prevent it from packing in the gullets of the teeth. If you don't do this, the saw will overheat and burn the wood. When you've cut through the stock, reach into the slots in the sides of the saw with a screwdriver to eject the waste plug.

3-7 When you must cut a hole to an odd diameter for which there is no multi-spur bit or hole saw, use a fly cutter. Be extremely careful with this tool. Fly cutters are not balanced, and they vibrate as they spin. This makes them notoriously "grabby" — they want to catch the stock. To protect yourself and your work, make doubly sure the work is secured to the table and the table is locked in place. Run the drill press as slow as it will go, and feed the cutter at a snail's pace. Keep your hands clear of the spinning cutter, and make sure it comes to a complete stop before you reach in to adjust the work or clear the waste. **Note:** Several mail-order companies offer fly cutters with two knives to balance each other. Although these tools run smoother and are less likely to catch, the same precautions still apply — secure the work, use a low speed, and feed the cutter very slowly.

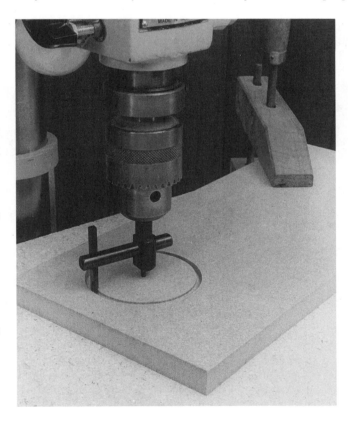

MATCHING HOLES

When you must drill matching holes on two or more separate workpieces, there are several methods you might use, depending on the number and the type of holes. The simplest method is to stack the parts and *pad drill* them — bore through the entire stack at once. (*SEE FIGURE 3-8.*) However, this works only when you're making through holes. To match stopped holes, make a holding fixture that fastens to the drill press table and locates each board in the same position under the bit. (*SEE FIGURE 3-9.*) Or make a drilling template and fasten it to each workpiece to guide the bit. (*SEE FIGURE 3-10.*)

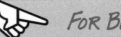

FOR BEST RESULTS

When using a holding fixture or a template, be especially careful to clean away the waste after drilling each workpiece. If the chips build up and pack against the holding surfaces of the jig, the jig won't hold the wood in the proper position and the holes won't match from piece to piece.

3-8 To bore matching holes through two or more parts, *pad drill* them. Stack the parts face to face and secure them with double-faced carpet tape. Lay out the holes on the top part, then drill through the entire stack. When you've finished drilling, take the stack apart and discard the tape.

(1)

(2)

3-9 When matching stopped holes or when the workpieces are too thick to pad drill, use a *holding fixture* to secure each part in the same position under the bit. There are many ways to make a holding fixture; two simple examples are shown. Cut a *notch jig* (1) to fit the shape of the parts and clamp it to the drill press table. Or, set up a *fence and stop* (2) on the table to hold one edge and one end of each part.

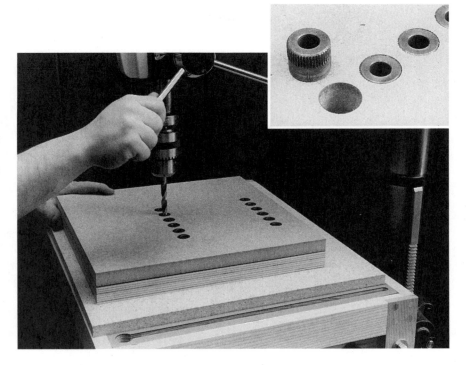

3-10 When you must drill the same sequence of holes over and over, it's often worth your while to make a single *drilling template* rather than set up several holding fixtures. Cut a plywood scrap to the same size and shape as the parts to be drilled, and install steel *drill bushings* of the proper diameters wherever you want to make a hole. (These bushings prevent the drill bit from enlarging the holes in the template.) Clamp the template to the workpiece or secure it with carpet tape. Then drill the holes, using the bushings to position and guide the bit. **Note:** Drill bushings are available from many mail-order woodworking suppliers.

EQUALLY SPACED HOLES

To drill a series of equally spaced holes, make a drilling template and attach it to the work. (*SEE FIGURE 3-11.*) Or make a special *pin stop* to space them. The advantage of the pin stop is that you can adjust the hole spacing to fit your needs. Attach an L-shaped stop to the fence to the left or right of the bit — the distance between the bit and the stop controls the spacing. Drill a hole, place a dowel pin in it, and slide it sideways until it butts against the stop. Drill a second hole and repeat until you have drilled the entire series. (*SEE FIGURES 3-12 AND 3-13.*) Refer to "Hole-Spacing Fixture" on page 56 for a more versatile pin-stop design.

3-11 You can make your own drilling template to drill equally spaced holes, or choose from several commercial templates. Most of these are designed to make holes in the sides of bookcases and cabinets for shelving support pins. The clear plastic template shown is designed to be used with a Vix bit, a device that automatically centers the bit in the template holes and limits the depth of the holes you drill.

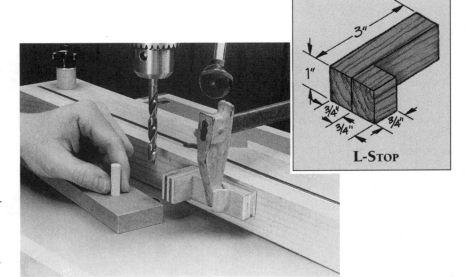

L-STOP

3-12 You can also use a *pin stop* to space holes. Clamp an L-shaped stop to a fence just above the work and to one side of the bit — the horizontal distance between the stop and the bit controls the spacing. Drill a hole and insert a dowel pin in it. Slide the work sideways until the pin butts against the stop.

3-13 Drill a second hole, holding the pin against the stop. Then take the pin out of the first hole and place it in the second. Slide the work sideways until the pin hits the stop, and drill a third hole. Repeat until you have drilled the entire series of holes. As long as the pin fits the holes with no play and the stop is fastened no more than 1/16 inch above the workpiece, the spacing between the holes will be uniform.

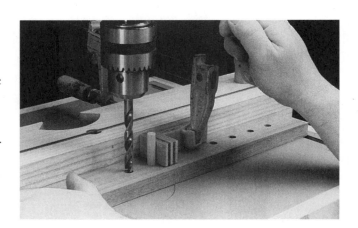

ANGLED HOLES

There are several ways to drill angled holes. The most common is to tilt the drill press table — front to back or side to side, however your table pivots. Use a protractor or a sliding T-bevel to set the angle, then fasten the fence on the *downhill* side of the bit — this will cradle the stock and prevent it from slipping. (*See Figures 3-14 and 3-15.*)

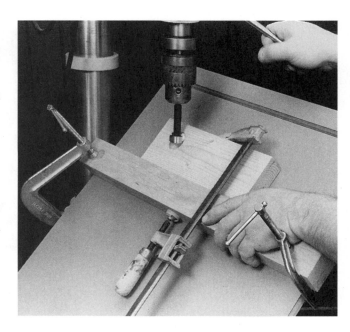

3-14 When drilling at an angle, tilt the drill press table. Mount a fence on the *downhill* side of the drill bit — otherwise, when you feed the bit, it will push the wood downhill. The fence acts as a stop to prevent this. **Note:** If a fence won't work, mount the stock in a vise, or clamp it to the table.

If you have an angle vise, you can use it to hold the stock at an angle instead of tilting the table. Or, if you have a radial drill press, tilt the head. (*SEE FIGURE 3-16.*)

3-15 When drilling at an angle, *feed the bit very slowly as you start the hole.* This is *extremely* important! Because the stock is at an angle, the sides of the bit will engage the wood before the lead point. If you feed the bit too quickly at the beginning, it will be deflected toward the downhill side of the table. The hole won't be positioned properly, nor will it be at the proper angle. Wait until the lead point is engaged and the entire tip of the bit is cutting before you feed at a normal rate.

TRY THIS TRICK

If you build a tilting table for your drill press, as shown in "Drill Press Table" on page 96, cut several wooden triangles to help set the angles you use often. Blunt the corners of each triangle, cutting them back about ¼ inch. To quickly tilt the table, select a triangle cut to the desired angle and slide it between the table and the base. When the underside of the table rests flat on the hypotenuse of the triangle, lock the supports.

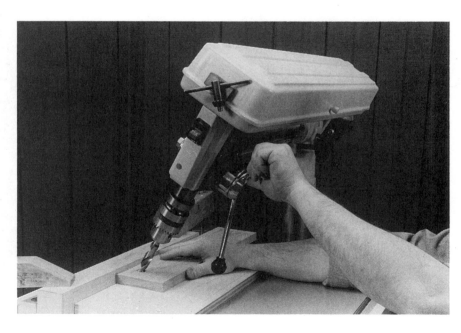

3-16 If you own a radial drill press, you must tilt the head to drill angled holes. To prevent the work from slipping, mount a fence or a stop to the downhill side of the bit, or clamp the work to the table.

HOLE-SPACING FIXTURE

If you regularly make cabinets, bookcases, and other projects in which you need to drill equally spaced holes, this fixture will come in handy. It's an adjustable *pin stop* (with a built-in pin) that mounts to a fence. As shown, it's designed to work with the fence in "Drill Press Table" on page 96, but you can adapt it to work with most fence designs.

1 **Set the distance between the** fence and the bit, then mount the fixture to the fence. Mark two parallel lines on a scrap, making the distance between them equal to the hole spacing. Center the bit over one line and drill a hole. Slide the scrap sideways and let the pin drop into the hole. Holding the scrap against the fence, center the second line under the bit. Tighten the wing nuts to lock the fixture down.

2 **To use the fixture, mark the** position of the first and last hole on the workpiece. Place the workpiece on the table, drill the first hole, and slide the workpiece sideways until the pin drops into the hole.

3 **Continue to press the work-** piece sideways *lightly* — this will take any play out of the jig. Drill the second hole, lift the pin out of the first hole, and slide the work sideways until the pin drops into the second hole. Repeat until you reach the layout mark for the last hole. If the last hole is off the mark, check your setup. **Note:** Run several test pieces before drilling good stock, to make sure your setup is accurate. Don't just drill two or three holes to check it out — drill the *entire series*.

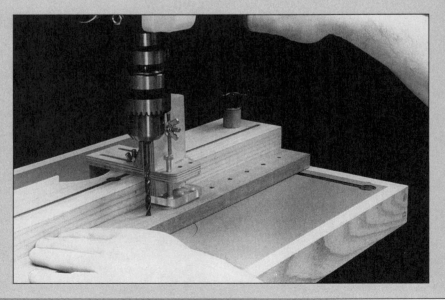

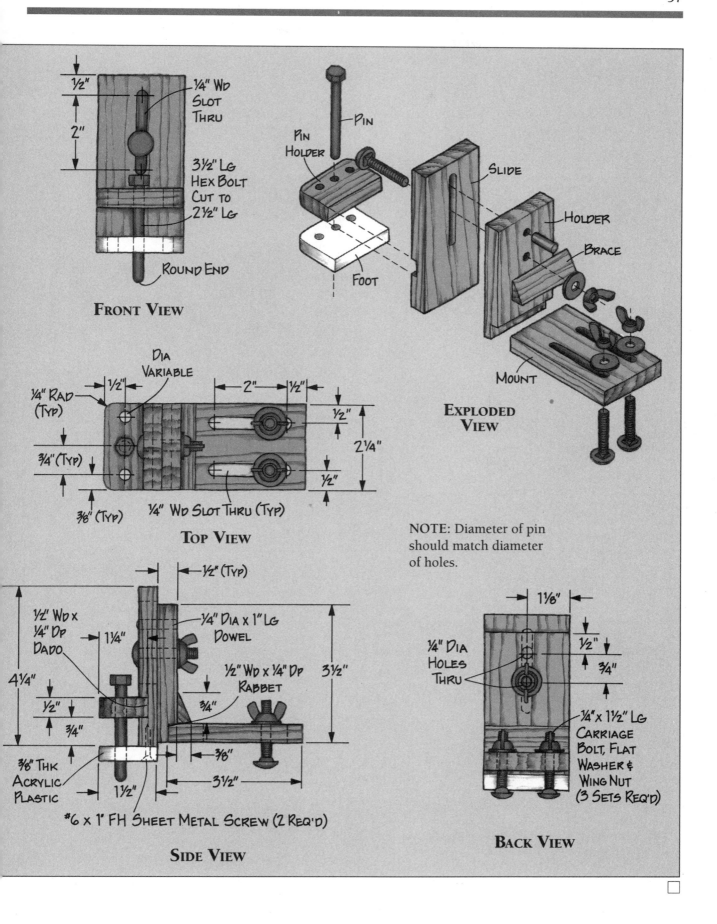

FRONT VIEW

½"
2"
¼" Wd Slot Thru
3½" Lg Hex Bolt Cut to 2½" Lg
Round End

TOP VIEW

Dia Variable
¼" Rad (Typ)
½"
2"
½"
½"
2¼"
¾" (Typ)
½"
⅜" (Typ)
¼" Wd Slot Thru (Typ)

EXPLODED VIEW

Pin
Pin Holder
Slide
Holder
Brace
Foot
Mount

NOTE: Diameter of pin should match diameter of holes.

SIDE VIEW

½" Wd x ¼" Dp Dado
1¼"
4¼"
½"
¾"
⅜" Thk Acrylic Plastic
1½"
½" (Typ)
¼" Dia x 1" Lg Dowel
½" Wd x ¼" Dp Rabbet
¾"
3½"
⅜"
3½"
#6 x 1" FH Sheet Metal Screw (2 Req'd)

BACK VIEW

1⅛"
½"
¾"
¼" Dia Holes Thru
¼" x 1½" Lg Carriage Bolt, Flat Washer & Wing Nut (3 Sets Req'd)

MAKING SCREW POCKETS

One of the most common applications for angled holes is making *screw pockets* — angled pockets and shank holes for screws that fasten the edge of one board to the face of another.

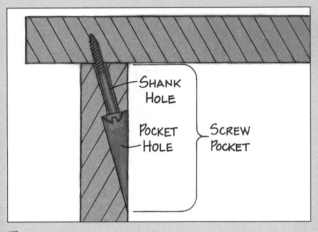

1 **Screw pockets are often used** in furniture construction to attach table aprons to tabletops, as shown. The screw passes through the *shank hole* in the apron and into the underside of the tabletop. The head rests in the counterbore or *pocket hole.* These holes are drilled at the same angle (usually 15 degrees) on the inside of the apron, where they won't be seen. The shank hole exits the edge of the apron about halfway between the faces.

2 **To drill a screw pocket, tilt** the drill press table to 15 degrees, attach a fence to the downhill side, and rest a scrap on the table to back up the work. To determine where to position the fence, divide the thickness of the board you're going to drill by 2. For example, if you're making a pocket hole in a ¾-inch-thick board, the calculation would be ¾ divided by 2, or ⅜. Position the fence ⅜ inch from the spot where the lead point of the drill bit touches the backup scrap. Position the workpiece under the bit with the edge on the scrap and the face against the fence. Clamp it to the fence to prevent it from shifting. Drill the pocket hole first, stopping about ⅜ inch from the backup scrap.

3 **Switch drill bits and make** the shank hole, drilling through the center of the pocket hole and into the scrap. **Note:** Craftsmen often make shank holes slightly larger in diameter than the shanks of the screws. This allows the tabletop to expand and contract with changes in relative humidity.

HOLES IN ROUND STOCK

When drilling round stock, cradle it to prevent it from rolling around on the drill press table. There are two ways to make a cradle — tilt the table and attach a fence to the downhill side, or make a *V-jig*. What-ever method you use, you must position the cradle so the corner or the V is directly under the tip of the bit. (*SEE FIGURES 3-17 AND 3-18.*)

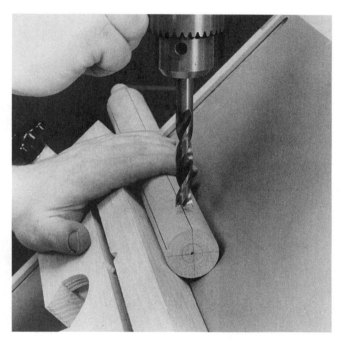

3-17 To hold round stock on the drill press, tilt the table to 45 degrees. Attach the fence to the downhill side to form a V-shaped cradle. Position the fence so the lead point on the bit is directly over the point of the V, where the fence meets the table. This ensures that the bit will pass through the center of the round stock.

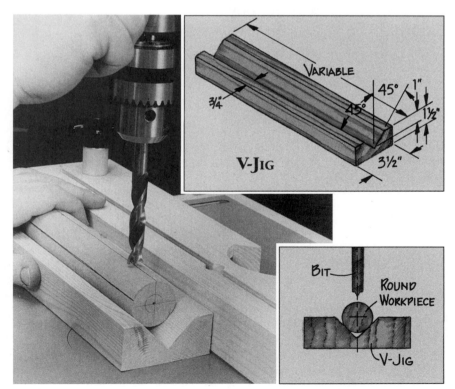

3-18 You can also use a V-jig to hold round stock for drilling. Clamp the jig to the drill press table so the point of the V is directly under the tip of the bit. **Note:** Cut V-jigs from two-by-four scraps, as shown. Use these as you would backup scraps — when they get chewed up, make new ones.

RADIAL HOLES

To arrange holes in an arc, all the same distance from a point on the workpiece, rotate the workpiece under the bit after you drill each hole. Use a nail or a screw as a pivot. (SEE FIGURE 3-19.)

CONCENTRIC HOLES

When you need to make a counterbore for a screw or create a rabbet around the circumference of a small hole, you can do so by drilling *concentric* holes. Drill two holes at the same mark, one larger than the other. (SEE FIGURE 3-20.)

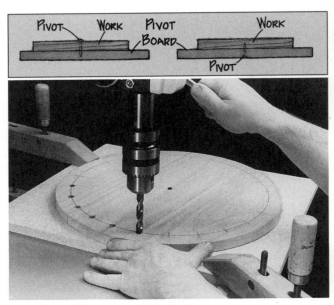

3-19 To drill a series of holes in an arc, pivot the workpiece under the bit. Drive a nail or a screw through the workpiece and into another board to serve as a pivot. Or, if you don't want a pivot hole in the workpiece, drive the nail or screw through a large scrap of plywood so that about ¼ inch of tip protrudes from the surface and let the work pivot on top of it. Position the work and the pivot board on the drill press table so the distance between the pivot and the center of the bit equals the radius of the arc. Lay out *radians* on the workpiece — lines that extend out from the pivot point through the arc where you plan to drill holes. Rotate the work around the pivot, drilling a hole where each radian crosses the arc.

TRY THIS TRICK

You can make rings and wheels with hole saws by cutting concentric holes. Cut the outside diameter of the wheel or ring, and clamp the disc in a hand screw clamp. (The wood grain should be parallel to the clamp jaws.) Then cut the inside, using the pilot hole to align the saw. Be careful not to overtighten the clamp or it will crush the wheel. To make thin wheels or rings, don't use a clamp. Instead stick the disc to a scrap board with double-faced carpet tape to cut the inside diameter.

3-20 To make a counterbore or create a rabbet around a hole, drill two *concentric* holes. Drill the larger hole first, stopping at a preset depth. Then drill the smaller hole, using the dimple left by the lead point of the larger bit to locate the smaller one. Or, if the drill bits have no point, clamp the workpiece to the drill press table so it can't move, drill the larger hole, change bits, and drill the smaller one.

HOLES FOR HARDWARE

Use the technique for drilling concentric holes to make counterbores and pockets for all kinds of fasteners. Drill a large hole to recess the head of a round-head screw, carriage bolt, or hex bolt, then make a smaller hole for the shank and threads. Seat flathead and ovalhead wood screws with a countersink, or make a counterbore, countersink, and shank hole at the same time with a screw drill. (SEE FIGURE 3-21.) Many types of screw drills are available from woodworking suppliers.

When drilling screw holes, you must sometimes make the shank hole in one workpiece the same size as the screw shank, and the matching *pilot hole* in the adjoining workpiece a little smaller so the threads bite into the wood. This lets you use the screws to draw the parts snug against one another. (SEE FIGURE 3-22.) When you must allow for wood movement, make the shank hole slightly oversize. Or, drill a slot for the screw shank, aligning the slot's long dimension with the direction of the movement. (SEE FIGURE 3-23.)

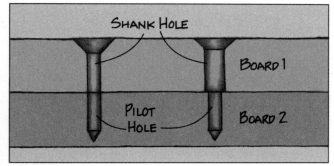

TRY THIS TRICK

To prevent a hex-head bolt from turning in its counterbore, make the counterbore the same diameter as the distance from flat to flat on the hex head. Then drive the bolt head into the counterbore or draw it in with a nut and washer, letting the corners bite into the wood.

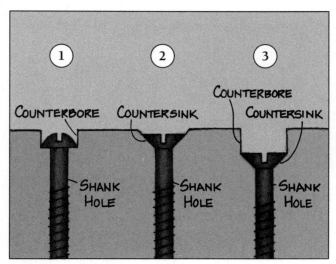

3-21 There are three types of recesses you can create for fasteners, each of them made by drilling concentric holes. A *counterbored hole* (1) will hold a roundhead screw, carriage bolt, or hex bolt. A *countersunk hole* (2) seats a flathead or ovalhead wood screw. And a hole that's counterbored *and* countersunk, often referred to as a *screw hole* (3), recesses a flathead screw beneath the wood surface, where it can be hidden with a wooden plug.

3-22 When assembling wooden parts with screws, decide whether you're going to use the fasteners simply to hold the parts together or whether you need to *draw* them together. Each requires slightly different screw holes. If you plan to clamp the parts together before you drive the screws, you can drill screw holes with shank holes the same diameter as the pilot holes. When you drive the screws, they will simply hold the parts in position. When you need the screws to provide clamping action and help draw the parts together, make the shaft holes larger than the pilot holes. The screw threads will bite into one part and snug it up to the other.

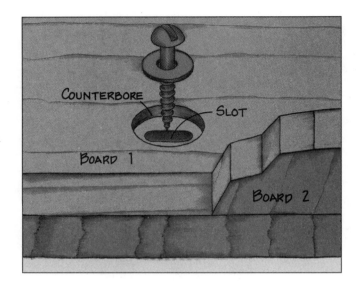

3-23 When you attach two
wooden parts with opposing wood
grain, you must allow the parts to
expand and contract separately. To
do this, fasten them together with
screws, but make the shank holes
slightly oversize. Or, instead of
drilling shank holes, make *slots* with
the long dimension parallel to the
wood movement. As the wood
expands and contracts, the screw
shanks will slide back and forth in
the oversize holes or slots. **Note:** To
make a slot, bore three overlapping
holes with the drill press, then
remove the waste between them
with a rat-tail file.

DRILLING END GRAIN

Drilling parallel to the wood grain is more work than
drilling across it. You're asking the drill bit to sever
the wood fibers instead of shave them, which requires
more power and creates more friction. The bit heats
up more quickly and is more likely to catch on the
stock or burn it. To get good results, you must adjust
your drilling technique to compensate.

When drilling end grain, the rules of thumb are:

■ *Slow down.* Because the bit is more likely to
overheat and grab the stock, use a slow speed and
moderate feed, and retract the bit often to clear chips.

■ *Get the right bit.* Although any bit will do in a
pinch, Forstner bits and multi-spur bits are not well
suited for end grain — they overheat and grab the
stock much too easily. Your best choices are brad-
point bits and spade bits. Twist bits and boring bits
are second choices.

■ *Support the wood.* As with any drilling operation, you must support and hold the wood as you
drill the end grain. Most drill presses, however, aren't
equipped to hold boards on end. You often have to
make a fixture to perform the operation safely and
accurately.

VERTICAL END DRILLING

When drilling the end of a small board, rest the opposite end on the drill press table. To hold the board
upright, make a *tall fence* and attach it to the drill
press table or the regular fence. (*SEE FIGURE 3-24.*) To
drill the end of a medium-size board, swing the table
out of the way and rest the opposite end on the base.

Tilt the table so the surface is vertical, and use it to
support the work. (*SEE FIGURE 3-25.*) Or, make a vertical table as shown in "Vertical Table" on page 64.

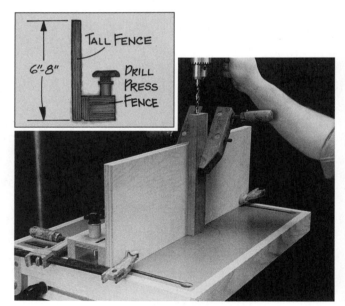

3-24 When drilling the ends of
small parts, use a *tall fence* to help
support them at the proper angle to
the bit. You can make a tall fence
from a scrap of plywood — cut it as
long as your regular drill press fence
and 6 to 8 inches wide. Fasten it to
your drill press fence with clamps,
bolts, or screws.

3-25 To drill the end of a board
that's too long to fit between the table
and the bit but that will fit between
the table and the base, swing the
table out of the way. Rest the opposite end of the board on the base,
using spacers as needed to raise the
board. Tilt the table so the work surface is vertical, and position it to support the face of the board. To make
sure the board doesn't shift as you
drill it, clamp it to the table.

3-26 Multipurpose tools such as
the Shopsmith and Total Shop con-
vert to horizontal boring machines,
allowing you unlimited work capacity. However, you must still prevent
the work from shifting as you feed
the bit. Either clamp the stock to
the table or back up one end with
a fence.

3-27 If you have a *drill guide* for a
portable drill and a *joint maker* or
horizontal routing jig for a router,
you can combine the two accessories
to make an excellent horizontal
boring jig. Simply attach the base of
the drill guide to the joint maker's
mounting plate where you would
ordinarily mount a router. Secure the
workpiece to the joint maker table,
and raise or lower the mounting
plate to adjust the vertical position of
the drill bit.

HORIZONTAL BORING

When the board is longer than the work capacity of
the drill press, it's easier to drill the end grain with the
board held horizontally. (This is called *horizontal
boring*.) There are three ways to bore the end of a horizontal board.

■ Use a horizontal boring machine, if you have
access to one. Many multipurpose woodworking tools
have a horizontal boring mode. (*See Figure 3-26.*)

■ Use a portable hand drill mounted in a drill
guide to bore the holes accurately. (*See Figure 3-27.*)

■ Make a template with a steel drill guide bushing
to guide your portable drill.

VERTICAL TABLE

This shop-made fixture fastens to the table shown in "Drill Press Table" on page 96, enabling you to drill the ends of medium-length boards that are shorter than the drill press work capacity.

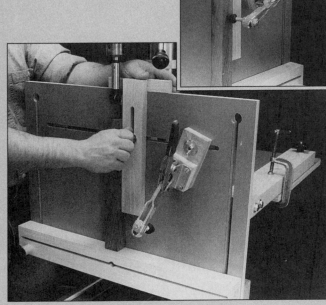

1 **To use the vertical table, first** rotate the regular drill press table 180 degrees around the column. Attach the supports to the table by inserting the heads of the mounting bolts in the fence-mounting slots. Slide the table forward or back to adjust the horizontal position of the work surface, then tighten the locking knobs and clamp the support arms to the table.

2 **Attach a fence to the vertical** slots to support the work at the proper height. Or, let the work rest on the drill press base. To help align the work parallel to the drill bit, attach a small T-square to the horizontal slot. To hold the work while you bore, mount the drill press clamp (shown in "Drill Press Table" on page 96) to either the horizontal or vertical slots.

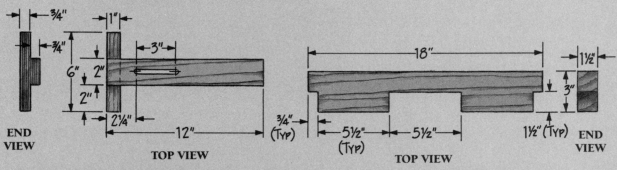

T-SQUARE DETAIL

END VIEW

TOP VIEW

¾"

¾"

1"

3"

6"

2"

2"

2¼"

12"

BRACE DETAIL

TOP VIEW

END VIEW

18"

1½"

3"

¾" (TYP)

5½" (TYP)

5½"

1½" (TYP)

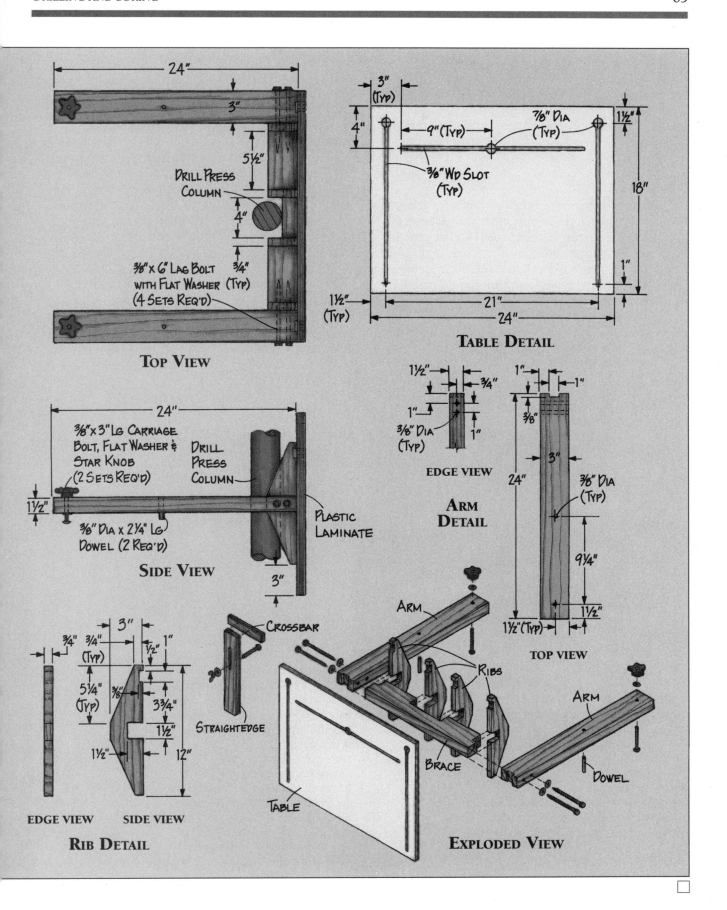

24"

3"

DRILL PRESS
COLUMN

5½"

4"

¾"

⅜" x 6" LAG BOLT
WITH FLAT WASHER (TYP)
(4 SETS REQ'D)

TOP VIEW

3"
(TYP)

4"

9" (TYP)

⅞" DIA
(TYP)

1½"

⅜" WD SLOT
(TYP)

18"

1"

1½"
(TYP)

21"

24"

TABLE DETAIL

24"

⅜" x 3" LG CARRIAGE
BOLT, FLAT WASHER &
STAR KNOB
(2 SETS REQ'D)

DRILL
PRESS
COLUMN

1½"

⅜" DIA x 2¼" LG
DOWEL (2 REQ'D)

PLASTIC
LAMINATE

3"

SIDE VIEW

1½"

¾"

1"

1"

⅜" DIA
(TYP)

EDGE VIEW

1"

1"

⅜"

3"

24"

⅜" DIA
(TYP)

9¼"

1½"

1½" (TYP)

TOP VIEW

**ARM
DETAIL**

¾"

¾"
(TYP)

3"

½"
(TYP)

1"

5¼"
(TYP)

⅜"

3¾"

1½"

1½"

12"

EDGE VIEW   **SIDE VIEW**

RIB DETAIL

CROSSBAR

STRAIGHTEDGE

ARM

RIBS

BRACE

TABLE

ARM

DOWEL

EXPLODED VIEW

DRILLING METAL AND PLASTIC

Occasionally, woodworkers drill holes in materials other than wood. Woodworking projects often require you to bore metal or plastic parts, and these require special care.

DRILLING METAL

Although there are many different types of metals and alloys, you can lump these all into two broad categories: *ferrous metals* (iron and steel) and *nonferrous metals* (brass, copper, aluminum, bronze, and so on). Ferrous metals are harder than nonferrous metals and are therefore more difficult to drill. But the methods for drilling them are much the same.

For either type of metal, begin by selecting a twist bit to make the hole. The vast majority of drill bits are designed for wood; only twist bits are made to bore metal. Mount the bit and set the drill press to a slow speed. Because metal is much harder and denser than wood, you cannot cut it as quickly. You can use slightly higher speeds when drilling nonferrous metals than when drilling iron and steel, but in general the speed should be slower than you'd use for wood. (If you're unsure of what speed to use, refer to the chart "Recommended Drill Press Speeds" on page 28.)

Lay out the locations of the holes on the metal by scratching the surface with a metal scribe. If the scratch marks are difficult to see, cover the surface with a dark surface dye, then scratch lines in the dye. (SEE FIGURE 3-28.) Where the lines intersect, mark the locations of the holes with a center punch. (SEE FIGURE 3-29.)

Position the work under the bit and secure it to the drill press before you start. Metal is more "grabby" than wood; the bits are more likely to catch as you work. And drilling is more dangerous when the bits do catch — a thin, whirling piece of metal becomes, in effect, a saw blade. To prevent this, clamp the work to the table or the fence, or hold it in a vise.

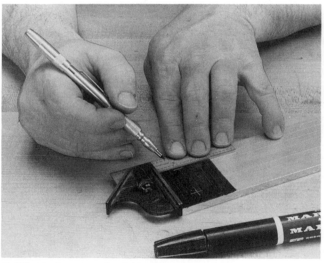

3-28 To lay out the hole locations on light-colored metal, first coat it with a surface dye such as copper sulfate — dyes are available from suppliers that sell machinist's tools and materials. Or, use a dark-colored broad-tipped marker instead. Make layout lines on the darkened surface with a scribe — the light metal will show through the dark color wherever you scratch it. After drilling the metal, wash away the color with water or the recommended solvent.

3-29 Where the lines intersect, mark each hole with a center punch, hitting the flat end of the punch with a cross peen hammer (shown), ball peen hammer, or another machinist's hammer. This will create a small indentation in the metal's surface. When accuracy is especially important, mark the holes with a prick punch first, tapping it with a hammer to make a tiny mark. Then enlarge this mark with a center punch. **Note:** Do *not* hit a hardened punch with a carpenter's hammer. The hammer head is tempered to make it tougher than nails (literally) and it may fly apart.

To help position the work precisely, mount a bit with a lead point in the chuck. Slide the mark for the hole directly under the point and clamp the work down. Change to a twist bit before you start drilling.

Put a few drops of oil around the indentation created by the center punch to lubricate the bit and keep it from overheating. (*SEE FIGURE 3-30.*) This isn't neces- sary when drilling nonferrous metals, but it's a must for iron and steel, and good practice in any event.

Feed the bit very slowly until the lips are fully engaged. Then increase the speed so the bit cuts a long, continuous metal ribbon. (*SEE FIGURE 3-31.*) When drilling large holes, start with a small bit, then enlarge the hole in several steps. Or, use a special *center drill* (available from machinist's suppliers). This helps ensure accuracy — it's easy for a large bit to wander off the mark left by a center punch. When you've finished the holes, remove the burrs around the edges with a countersink. (*SEE FIGURE 3-32.*)

3-30 Heat is the enemy of all drill bits — as the cutting edges on the lips heat up, they dull more quickly. If they overheat, the temper of the steel is reduced and it may no longer hold a sharp edge. To keep the bit running as cool as possible, lubricate the cutting edges. Place a few drops of oil on the metal surface before you begin the hole, then add a drop or two every few minutes as you continue to drill.

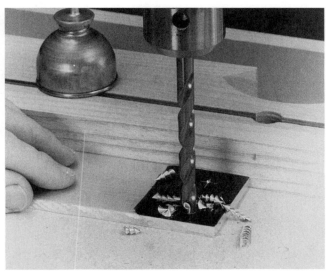

3-31 Feed the drill very slowly at first, letting the lips enlarge the indentation made by the center punch. When the entire tip of the drill is cutting the metal, increase the feed pressure until the drill is cutting a long, thin ribbon of metal.

3-32 When you've drilled through the metal, there will likely be small burrs around the circumferences of the entrance and the exit holes. Remove these with a few quick twists of a hand countersink.

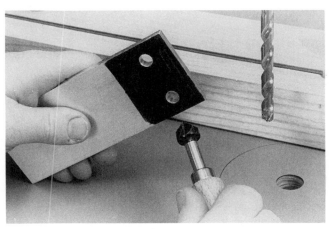

DRILLING PLASTIC

Twist bits also work best when drilling plastics such as acrylic (Plexiglas) and polycarbonate (Lexan), although these materials are soft enough to use other common cutters such as brad-point bits, Forstner bits, and multi-spur bits. There are also special plastic-drilling bits available — twist bits with the tips ground to 60 degrees. If you work with plastic often, you may want to invest in a set of these.

Set the drill press to run a little slower than you would for wood, but a little faster than for metal — refer to the chart "Recommended Drill Press Speeds" on page 28. Lay out the holes on the paper covering with a sharp pencil. (Most plastic comes with protective paper on both faces — leave this in place until you've machined the workpiece.) Mark the hole locations with an awl or a prick punch. (SEE FIGURE 3-33.)

Place a scrap of plywood or hardwood on the drill press table to back up the plastic, then position the plastic under the bit. If you're drilling a large hole or

a small part, secure the plastic to the drill press table or fence. Place a few drops of oil or kerosene under the tip of the bit to help lubricate the cutting edges, and begin drilling. Feed the bit slowly until you're sure it's started in the proper location, then increase the feed pressure. (SEE FIGURE 3-34.) When the bit exits the hole, clean up the burrs and splinters around the edges with a hand countersink.

A BIT OF ADVICE

Finding and holding the correct feed pressure is the most important (and the trickiest) step in drilling plastic. If you apply too much feed pressure, the plastic may chip or break. If you apply too little, the bit will dwell in the hole too long, causing it to heat up and melt the plastic.

3-33 Lay out the holes on the pro-tective paper that covers the plastic. Or, if the plastic isn't covered, affix a sheet of paper to its surface with spray adhesive. Mark the locations of the holes with an awl or a prick punch, twisting the point into the plastic to create a small indentation.

3-34 Before you start drilling, put a few drops of lubricant on the plastic surface in the vicinity of the hole to keep the bit running cool. (If it heats up, the plastic will melt.) Start the hole slowly, then increase the feed pressure until thin shavings or a long ribbon of plastic is ejected from the hole. Make sure your workpiece is backed up with a wood scrap — when the bit exits, plastic will fracture and chip if it isn't well supported.

4

DRILL PRESS JOINERY

Y ou can cut a good deal of
joinery on a drill press, including
one of the strongest and most useful joints there is — the mortise-
and-tenon joint. This assembly
consists of two parts — one end of
a workpiece forms a tenon, which
fits into a mortise in an adjoining
piece. The mortise is just a hole,
sized and shaped to fit the tenon.
You can cut both round and square
mortises on a drill press.

Additionally, you can cut holes
for pegs and dowels. Driving
square pegs into round holes was
once common practice for assembling frame-and-panel construc-
tions, and craftsmen still resort to
this technique when reproducing
classic designs. You are more likely,
however, to use dowel joints. The
dowels can be thought of as small,
round "loose" tenons that fit
matching pairs of round mortises,
joining frames and other assemblies. You can also use dowels and
pegs for decorative joinery.

SQUARE MORTISES AND TENONS

There are two ways to make a square mortise on a drill press. You can drill a round hole to remove most of the waste, then square the corners with a chisel; or you can use a hollow chisel mortising attachment to remove the waste and cut the corners square in one step.

USING HAND CHISELS

To make a square mortise with a drill press, lay out the mortise on the stock. (*See Figure 4-1.*) Choose a drill bit that's the same diameter as the mortise is wide, and drill a row of overlapping holes. Use a fence to guide the workpiece so the row will be perfectly straight. (*See Figure 4-2.*) Square the ends of the mortise and shave the sides flat with hand chisels. A mortising chisel works best for cutting end grain; a bevel-edge chisel or a paring chisel is better for the sides. (*See Figure 4-3.*) If you need to cut the mortise a little deeper or make the bottom perfectly flat, use a router plane. (*See Figure 4-4.*)

Cut the tenon to fit the mortise. (*See Figure 4-5.*) If necessary, use a shoulder plane or a bullnose plane to shave the cheeks and shoulders of the tenon to get the fit you want. (*See Figure 4-6.*)

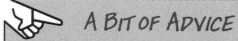

A Bit of Advice

Cut tenons about 1/16 inch shorter than the mortises are deep to allow room for glue.

4-2 Remove most of the waste from the mortise by drilling a row of overlapping holes. Select a drill bit with a lead point that leaves a hole with a relatively flat bottom — a brad-point bit works well. Set the depth stop to drill the holes as deep as you want to make the mortise. Use a fence to help position the stock under the bit — this and the centerline will help you drill the holes in a straight row. Drill the two end holes first, then work your way from one end to the other, overlapping the holes by no more than one-quarter of their diameter. **Note:** If you drill the holes too close together, the drill bit will drift in the stock.

4-1 When cutting a square mortise-and-tenon joint on a drill press, use a marking gauge to lay out the mortise. Scribe lines to define the sides and ends of the mortise, then make a deep line parallel to the side lines and centered between them. This centerline will help line up the lead point and start the drill bit.

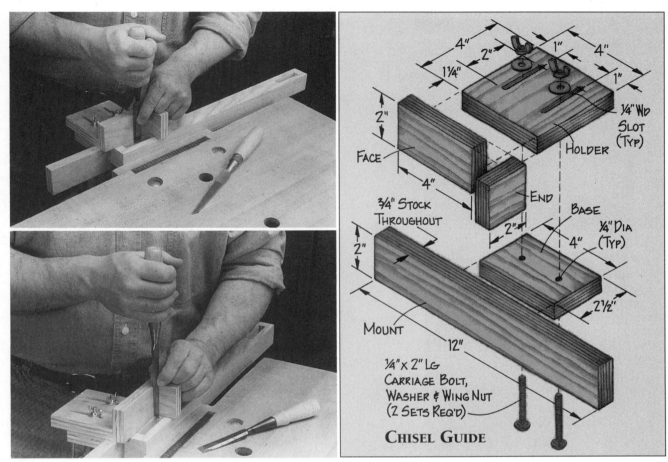

CHISEL GUIDE

4-3 Square the ends and shave the sides of the mortise with hand chisels, removing the rest of the waste. To keep the sides and ends perpendicular to the surface and perfectly straight, clamp a block of wood to the surface of the workpiece and use it to guide the back of the chisel. Or, make the simple chisel guide shown. Use a mortising chisel to cut the end grain — the long cutting edge and thick body are designed to sever the wood fibers cleanly and split them away from the stock in one step. Shave the sides with a wide bevel-edge chisel or a paring chisel — the broad cutting edge removes stock quickly and helps keep the cut surfaces flat.

4-4 If you need a mortise with a perfectly flat bottom, or if you must cut it just a little deeper, shave the bottom with a small router plane. The plane has an L-shaped iron that reaches down into the recess.

4-5 Cut a square tenon by making one or more rabbets in the end of the stock, on the faces and the edges. The bottoms of the rabbets become the *cheeks* of the tenon, while sides become the *shoulders*. You can cut a tenon on a table saw, using a dado cutter or saw blade (shown), or on a table-mounted router, using a straight bit.

4-6 For final fitting, trim the cheeks and shoulders of the tenon with a shoulder plane. (You can also use a bullnose plane.) If you plan to glue the tenon in the mortise, the best fit is a *slip fit*. The tenon should slip into the mortise without having to be forced, the shoulders should sit tight against the adjoining surface, and there should be very little play.

USING A HOLLOW CHISEL

You can also use a hollow chisel mortising accessory to make square mortises. If you do a lot of mortising, this fixture will save you time. Setup, however, is extremely important — the hollow chisel and bit must be properly aligned and adjusted; otherwise the fixture will not cut a square hole and you may damage the tool.

Select a hollow chisel and a bit the same size as the width of the mortise. If the mortise is wider than the available chisels, pick a chisel that's just a little over half the width of the mortise. Attach the chisel holder to the end of the quill and mount the hollow chisel in it. Square one side of the chisel to the drill press fence. (*See Figure 4-7.*) Slide the bit up through the chisel and mount it in the chuck so the spurs on the end of the bit clear the end of the chisel by $\frac{1}{32}$ to $\frac{1}{16}$ inch. (*See Figure 4-8.*) Attach the hold-down to the fence and adjust it to keep the work on the table.

4-7 To set up a hollow chisel mortiser, attach the chisel holder to the end of the quill and mount the chisel in the holder. Position the drill press fence to guide the work, then check that the side of the chisel is square to the fence. If necessary, loosen the mounting screw and turn the chisel until it's properly positioned. If the chisel and the fence aren't square to one another, the sides of the mortise won't be straight. **Note:** Mount the chisel so that a waste port faces you. This will help keep the mortise from filling up with chips. It will also make it easier to brush or blow away the waste.

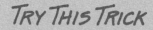

Try This Trick

Spray the inside of your hollow chisels with silicone to reduce friction, keep the bits running cooler, and help evacuate chips.

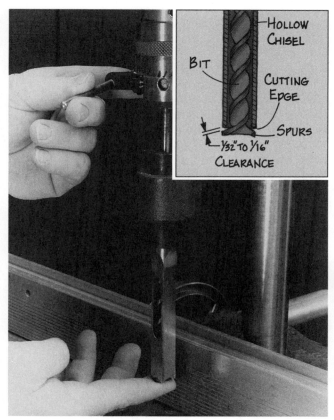

Adjust the drill press speed between 1,200 and 1,800 rpm. (The larger the chisel, the slower the speed.) Position the workpiece under the chisel and use the quill feed to plunge the chisel into the wood. (*See Figure 4-9.*) After making the first cut, turn off the machine and check the clearance between the chisel and the bit. If it has changed, the bit isn't properly seated in the chuck; reposition the bit and try again.

Bore rows of overlapping square holes to cut the mortise. Start with the two end holes, then work your way from one end to the other. If you must cut two or more rows to form the mortise, make the corner holes in the first row, then remove the waste between them. Repeat for the second row. (*See Figure 4-10.*)

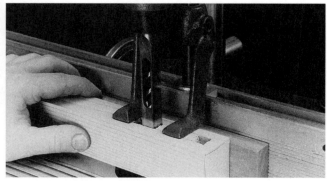

4-8 Slide the bit up inside the chisel and into the chuck. Adjust the clearance between the spurs on the end of the bit and the end of the chisel. This must be no less than $\frac{1}{32}$ inch and no more than $\frac{1}{16}$ inch. *This is very important!* If the clearance is too little, the bit will rub on the chisel. The resulting friction will heat the chisel and the bit, ruining both. If it's too large, the spurs won't break up the wood chips and they will clog the chisel. This, too, will result in overheating.

4-9 Attach the hold-down to the fence and adjust it to keep the stock on the table. Turn on the drill press and feed the chisel slowly with firm pressure. Don't try to feed too fast; give the bit plenty of time to evacuate the chips. It also helps to retract the chisel frequently during each cut — plunge the chisel into the wood, hold the pressure for a few seconds, retract the chisel, and repeat. (This technique is especially useful when mortising hardwoods.)

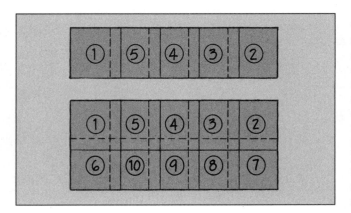

4-10 Form the mortise by drilling rows of overlapping holes. When drilling just a single row, make the end holes first, then work your way from one end to the other. When drilling a double row, make the corner holes in the first row, then remove the waste in between. Repeat for each succeeding row. **Note:** Ideally, the holes should overlap no more than one-quarter of the width of the chisel. Otherwise, the chisel may drift in the cut.

CHISEL BIT

Before the invention of the hollow chisel, old-time mortising machines were made like drill presses but were fitted with a mortising chisel instead of a bit. To use these machines, craftsmen drilled out most of the waste from a mortise, then cut away the rest with the chisel. The chisel was rotated by hand between cuts to shave the sides and ends of the mortise, and the machine kept the chisel straight up and down so the cut surfaces were flat and square.

You can duplicate this machine by making a *chisel bit* from a mortising chisel. Remove the handle from a ¼-inch mortising chisel, and cut off the tang or socket. Grind or file a flat on a ½-inch-diameter, 4-inch-long steel rod, and attach the *back* of the chisel to the flat with silver solder. Then make a three-piece wooden clamp that fits over the quill and the chuck sleeve to prevent the chuck from turning — this is a nonpowered operation.

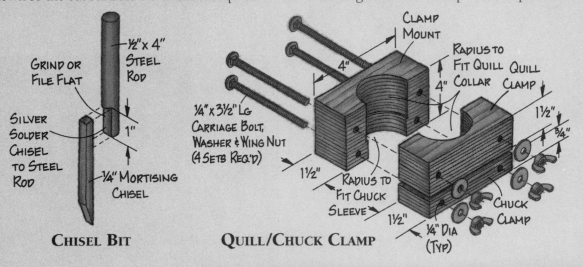

CHISEL BIT   **QUILL/CHUCK CLAMP**

1 **To make a mortise, first** drill out most of the waste from the workpiece. Mount the chisel bit in the chuck and turn it so the cutting edge is roughly parallel to the front edge of the drill press table. Secure the clamp to the chuck sleeve and the quill collar to keep the chuck from turning. Set the depth stop to halt the chisel when the cutting edge reaches the bottom of the mortise. Place the workpiece under the chisel so the layout line for the side of the mortise lines up with the cutting edge. Using the quill feed, slowly advance the chisel, shaving the side of the mortise. Retract the chisel, move the workpiece sideways slightly, and repeat. Continue until you have trued the entire side of the mortise. Turn the stock end for end and shave the opposite side.

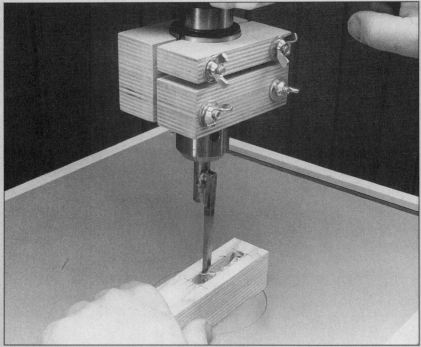

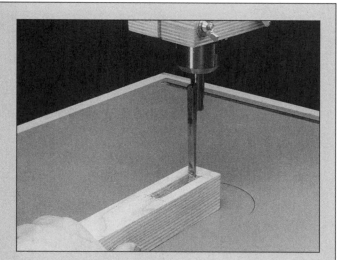

2 **Loosen the portion of the** wood clamp that holds the chuck. (Be careful not to loosen the part that is secured to the quill.) Turn the chuck 90 degrees so the cutting edge is roughly parallel to the side of the drill press table. Tighten the wooden clamp and shave the ends of the mortise, squaring the corners as you do so.

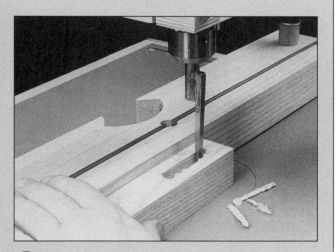

3 **To make the sides of the** mortises perfectly flat and parallel to the length of the stock, use the fence to guide the workpiece. Position the fence parallel to the cutting edge of the chisel and the necessary distance away from it. Move the stock sideways between each cut, keeping it pressed against the fence.

ROUND MORTISES AND TENONS

Making a round mortise is little different than drilling a hole. There are some important tricks, however, especially when making the mortise in round stock.

DRILLING A ROUND MORTISE

To mortise round stock, rest it in a V-jig and hold the jig against a fence as you work. Remember to position the fence so the lead point of the bit is directly over the point of the V.

When mortising round chair legs and other turned furniture parts, sometimes you must drill a row of mortises parallel to one another along the length of the workpiece. To do this, join two strips of wood to make a long, V-shaped straightedge. Rest this straightedge on the cylinder, and use it to draw a layout line parallel to the axis. (*SEE FIGURE 4-11.*) Mark the positions of the mortises along this line and drill them. (*SEE FIGURE 4-12.*)

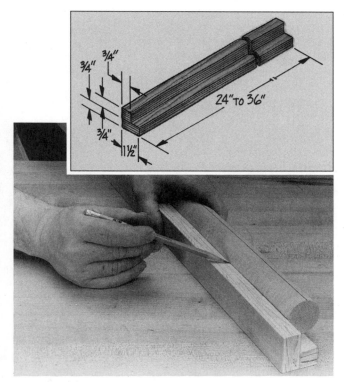

4-11 To mark round stock parallel to its axis, make a V-shaped straightedge from two strips of wood, as shown. Place this straightedge on the stock so the cylinder or the turning rests in the V, then use one of the long edges as a guide to mark the wood.

4-12 To drill a row of mortises in round stock, mark a line down its length. Measure the location of each hole along the line. Set up the drill press fence to guide a V-jig so the bit is centered over the point of the V. Place the stock in the V-jig, rotate it until the line is directly under the bit, and drill the mortises at the marks. The row of mortises will be perfectly straight and each mortise will be parallel to the others.

4-13 To drill two mortises — or two rows of mortises — at a precise angle to one another, draw two lines that intersect at the desired angle on a piece of stiff paper or cardboard. Attach this layout to one end of the round stock, driving a tack or a nail through the paper where the lines intersect and into the center of the cylinder.

4-14 Transfer the angle on the paper to the surface of the round stock by making marks where the lines on the paper cross the circumference. Using a V-shaped straightedge, make a line down the length of the stock at each mark. Measure the locations of the mortises along the lines and drill them, using a V-jig and a fence to hold and position the stock.

To drill two holes (or two rows of holes) at a precise angle to one another, use a protractor to lay out the angles on a small piece of cardboard or stiff paper. Fasten this to the end of the cylinder, and mark where the lines meet the circumference of the stock. (*SEE FIGURES 4-13 AND 4-14.*) Then use the V-shaped straightedge to draw a line at each mark. Measure the positions of the mortises along these lines and drill them.

 FOR BEST RESULTS

If you must make round mortises in a turning with a complex shape, first turn the stock to a simple cylinder. Mark this cylinder and drill the necessary holes in it, then turn the shape.

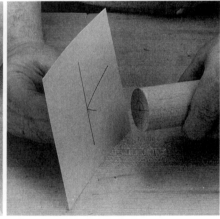
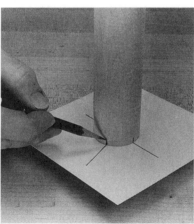

CUTTING A ROUND TENON

You can also make round tenons on your drill press. Use a tenon cutter to cut them to a precise diameter, holding the stock vertically. Set the drill press to run at a medium-slow speed (1,200 to 1,800 rpm). (*See* Figure 4-15.) Or, you can cut the tenons on a lathe, using calipers to gauge their exact size. (*See* Figure 4-16.)

4-15 Tenon cutters create round tenons of precise diameters in the ends of both round and square workpieces. Use these accessories in a horizontal boring machine, if one is available to you. If not, set up a drill press to hold the stock vertically, either by tilting the table 90 degrees or by using the vertical table fixture shown in "Vertical Table" on page 64. If the stock is round, mount it in a V-jig.

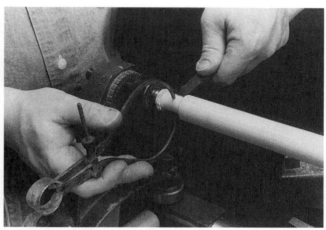

4-16 You can also turn tenons on a lathe, using calipers to gauge the diameter. Set the opening between the caliper arms to the desired diameter, then turn the tenon until the calipers slip over the tenon. Some craftsmen set the calipers $\frac{1}{32}$ to $\frac{1}{16}$ inch greater than the desired diameter, then finish turning the tenon with a file until they get the fit they want.

TRY THIS TRICK

When turning round tenons, make *fixed calipers* from a scrap of plywood to simplify the task. Drill a hole the same size as the tenon you want to turn, then open up this hole to the edge of the plywood. Cut a $\frac{1}{16}$-inch step in one edge of the opening, making the mouth $\frac{1}{16}$ inch wider than the desired diameter. Turn the tenon aggressively until the calipers slip over the stock up to the step. After that, remove stock more slowly. You can even drill a test hole in the calipers to help size the tenon precisely.

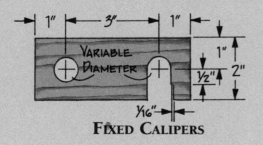

FIXED CALIPERS

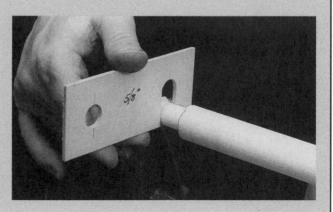

DOWELS AND PEGS

Dowels and pegs look somewhat the same, but they are used differently. Dowels are "loose" tenons that join two boards. The dowels rest in stopped holes, straddling the seam between the parts. They help position the boards during assembly and reinforce the joint afterwards. Pegs, on the other hand, are wooden nails. (The old-time term for them is *trenails* or "tree-nails.") They are driven through two boards, pinning them together. To install either dowels or pegs, you must drill matching holes in the adjoining boards.

DOWELING

Dowels often join boards edge to edge. To make the matching holes in the edges, use a *doweling jig* to guide the drill bit. If the board is not very wide, bore the

edge on the drill press, using a fence to help support it. (*SEE FIGURE 4-17.*) If it's wider than you can handle comfortably, use a portable drill to make the holes.

When doweling the ends and faces of boards, it's often easier to dispense with the doweling jig. There are four common methods for making matching dowel holes; use whichever one best suits the project:

■ Carefully mark the holes on the separate parts, making sure that the space between them is the same on each. It helps to make a storystick or a layout template to mark each part exactly the same. (*SEE FIGURE 4-18.*)

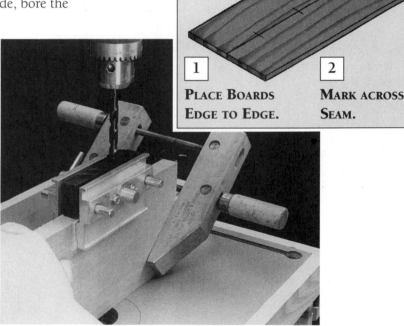

1 PLACE BOARDS EDGE TO EDGE.   2 MARK ACROSS SEAM.

4-17 **When joining boards edge to** edge, drill dowel holes with the aid of a *doweling jig.* Butt the adjoining edges of the boards together, and make a mark across the seam wherever you want to locate a dowel. Use these marks to position the doweling jig. Clamp the jig to the edge of the board at a mark, then drill a stopped hole in the edge, using the jig to guide the bit. Repeat at each mark.

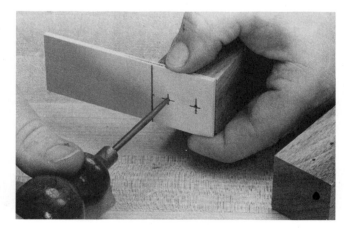

4-18 **To lay out matching holes on** two adjoining parts, make a layout template from a scrap of thin hardboard or plywood. Carefully measure and mark the positions of the holes on the template, then drill $1/16$-inch-diameter holes through the template at each mark. Clamp the template to the wood part. With an awl or a prick punch, poke through the holes in the template, making indentations in the wood surface. Use these indentations to locate the bit when drilling the holes.

■ Use the drill press fence and stops to position each part exactly the same on the table, or make a positioning jig and clamp it to the table. (*See Figure 4-19.*)

■ Lay out and drill the holes in one part, then insert *dowel centers* in the holes you just drilled. Use these centers to mark the hole locations on the adjoining part. (*See Figure 4-20.*)

■ Make a drilling template with steel drill bushings and fasten it to each part before drilling the holes. Use the template to position the bit for each hole. (*See Figure 4-21.*)

After drilling, dry assemble the parts (that is, assemble them without glue) to make sure the holes match properly. (*See Figure 4-22.*) When you're satisfied that they do, apply glue to the adjoining surfaces of the boards and the inside surfaces of the holes. Insert the dowels in one part, then clamp the parts together.

Where to Find It

You can purchase steel drill bushings of various sizes from:

Woodcraft Supply Corp.
210 Wood County Industrial Park
Parkersburg, WV 26102

Woodworker's Supply
1108 North Glenn Road
Casper, WY 82601

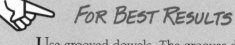
For Best Results

Use grooved dowels. The grooves distribute the glue evenly around the surfaces of the dowels and relieve the hydraulic pressure of the glue that's trapped in the holes. If this pressure is not dissipated, it may prevent the boards from joining properly.

STRAIGHT GROOVE   **SPIRAL GROOVE**

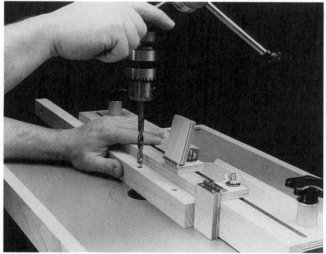
4-19 When drilling more than one dowel hole in a line, attach *flip-up stops* to the fence, as shown. Use one stop for every hole you want to drill. Hold the stock against the fence and push it against the first stop to position it for the first hole. Drill the hole, flip the stop out of the way, and slide the board sideways until it hits another stop. Repeat until you have drilled all the holes.

4-20 Mark matching dowel holes on two dissimilar parts with *dowel centers*. Drill dowel holes in one part and insert centers in the holes. Position the parts as you will join them, and press them together with enough force that the points on the dowel centers leave indentations in the surface of the adjoining part. Drill matching holes at the indentations.

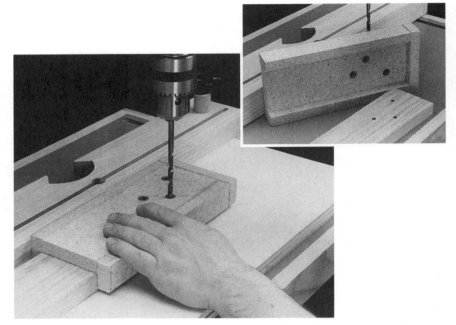

4-21 When you must drill holes in a complex pattern and it's not convenient to use dowel centers, make a *drilling template.* Cut a piece of plywood or particleboard to fit over the part, and install a steel drill bushing at each hole location — just drill a hole the same size as the barrel of the drill bushing and drive it into the template with a ball peen or cross peen hammer. To help position the template accurately, attach stops or fences around the perimeter of the template. Place the template over the workpiece and use the metal bushings to position the bit.

4-22 Fit the dowel joint together before gluing it. It should slip together like a well-made mortise-and-tenon joint; it needn't be a tight fit. More often than not, one or more sets of dowel holes will be mismatched. When this is the case, you can either plug them and redrill them, or shave the side of a dowel to get it to fit the off-center hole. The joint won't be quite as strong as it might have been, but as long as at least two sets of holes line up correctly, it should still offer adequate strength.

INSTALLING PEGS

Pegs are often used to lock wooden joints together. Tabletops, for example, sometimes have caps or *breadboards* on the ends to hide the end grain. The edges of the breadboards are grooved to fit over tongues on the ends of the tabletop. Cabinetmakers drive pegs through these tongue-and-groove joints to lock the parts together. You can also peg the mortise-and-tenon joints that hold door frames together, as well as bridle joints, lap joints, and many others.

Pegs are easier to install than dowels. Simply put the assembly together, drill the pilot holes through both parts, and drive the pegs into the holes. Typically, it's best to use pegs that are gently tapered or slightly larger than the pilot holes. This helps the pegs stay put. You might also drive square pegs into round holes — this geometry keeps the pegs from working loose. (*SEE FIGURES 4-23 THROUGH 4-26.*)

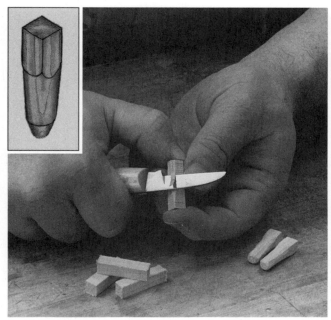

4-23 Cut square pegs from an extremely hard wood, such as rock maple or hickory. Whittle each peg, making it more and more round toward one end. The square portion should be just ¼ to ⅜ inch long.

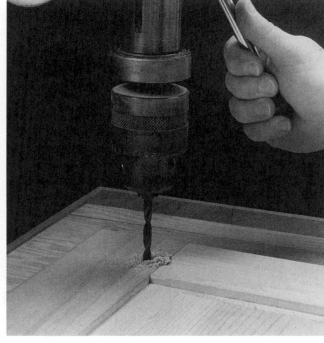

4-24 Assemble the joint and drill peg holes through both parts. The diameter of the holes should match the width of the square pegs.

4-25 Coat each peg with glue and drive it into a hole, round end first, with a mallet. Tap it in so the top is flush with the surface and the square portion is wedged in the hole. Be careful not to hit the peg so hard that it splinters.

4-26 If the rounded portions of the pegs protrude from the back of the assembly, cut them flush with the wood surface and sand them. **Note:** As wooden furniture expands and contracts over a century or more, it begins to push the pegs out of their holes. If you're making reproduction furniture, you may want to drive the pegs so the *square* portions protrude about ¹⁄₁₆ inch. This will help make the project look old.

5

SPECIAL DRILL PRESS TECHNIQUES

With the proper jigs and accessories, your drill press will do much more than make holes and mortises.

Add a shank-mounted drum sander and a cutout table, for example, and the drill press becomes a sanding machine. Use this setup to true sawed edges, smooth contoured surfaces, and sand boards to precise dimensions.

With a cutout fence, you can use the drill press as an overhead router to duplicate patterns, make signs, rout turned parts, and cut molded shapes, as well as perform many simple routing operations. Or, mount a small grindstone in the chuck and make a tool holder, and the drill press becomes a sharpener for chisels, knives, scrapers, and other metal tools.

These are just a few of the more common non-drilling applications. You can also use this tool as a grinder, shaper, and lathe. Some craftsmen even use their press as an emergency clamp, squeezing the work between the chuck and the table! With a little imagination, you can perform dozens of additional woodworking operations on the drill press.

SANDING

There are many sizes of shank-mounted sanding drums available for the drill press, from ½ inch to 3 inches in diameter. Typically, the sanding surfaces on these drums are backed by hard rubber, making it possible to sand flat surfaces and square corners.

To use these sanding drums, however, you must either modify your drill press table or make a special sanding surface for the machine. The work surface should have a hole directly beneath the chuck. When mounted, the bottom of the sanding drum should project down into the hole, to allow you access to the abrasive with your workpiece. (*SEE FIGURES 5-1 AND 5-2.*) **Note:** The "Drill Press Table" shown on page 96 has a cutout in the work surface that accepts collars with different inside diameters. This allows you to work with different sizes of drum sanders.

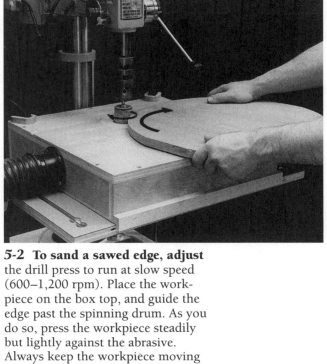

5-2 To sand a sawed edge, adjust the drill press to run at slow speed (600–1,200 rpm). Place the workpiece on the box top, and guide the edge past the spinning drum. As you do so, press the workpiece steadily but lightly against the abrasive. Always keep the workpiece moving *opposite* the direction of rotation. Don't press too hard; the abrasive will load up with sawdust and the wood will burn.

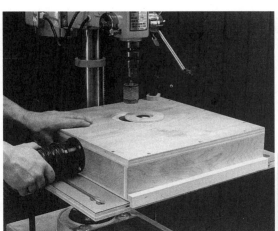

5-1 To use a sanding drum on a drill press, make a box to fit on your drill press table. Cut out the top of the box to make a hole slightly larger than the diameter of the drum sander, and fasten the box to the drill press table so the hole is directly beneath the chuck. Mount the drum and extend the quill until the bottom of the drum is below the top surface of the box. **Note:** The box shown has interchangeable collars so you can use different sizes of drum sanders, and a dust collection port on the side to let you collect sawdust as you sand. The cleat on the front helps position the box so the cutout is directly below the chuck.

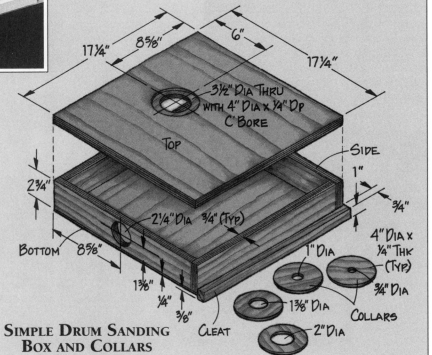

SIMPLE DRUM SANDING BOX AND COLLARS

You can also use a drum sander with a fence to sand boards to precise widths and thicknesses. Place the fence a specific distance from the sanding drum, then feed the wood between the fence and the drum.

(SEE FIGURES 5-3 AND 5-4.) Or, sand exact patterns by making a disc the same diameter as the sanding drum and attaching it to the table directly under this drum. Use the disc as a guide for a template. (SEE FIGURE 5-5.)

5-3 To sand a small board to a specific thickness, secure a fence to the drum sanding box or cutout table. Adjust the fence position about $\frac{1}{64}$ to $\frac{1}{32}$ inch closer to the sanding drum than the board is thick. Turn on the drill press, and feed the board between the fence and the sanding drum, *against* the direction of rotation. When you've completed the pass, reposition the fence $\frac{1}{64}$ to $\frac{1}{32}$ inch closer and repeat. Continue removing a fraction of an inch of stock at a time until you arrive at the thickness you want. **Note:** Don't try to remove too much stock in a single pass. The abrasive will load with sawdust, it will become harder to feed the board, and the risk of kickback will increase.

5-4 You can also use a fence and a sanding drum to sand boards to a specific width or to straighten an edge. This is a good way to "joint" pieces of plywood. If you attempt to run plywood across an ordinary jointer, the glue between the plies will abrade the knives, digging small ridges in them. Plywood can't harm the abrasive on a sanding drum, however. As long as the fence on the drill press is straight and sufficiently long, this setup will do an excellent job of edge-jointing while sparing your jointer knives.

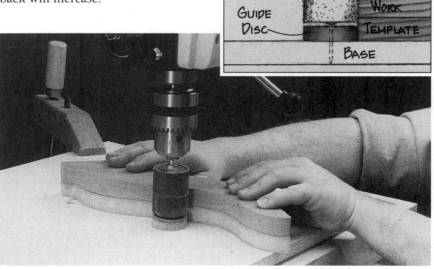

5-5 To sand a precise shape, cut a wooden disc the same diameter as the sanding drum. Attach this disc to a sheet of plywood or particleboard, and clamp the sheet to the drill press table so the disc is directly beneath the drum. Make a template for the shape you wish to sand from stock that's about $\frac{1}{4}$ inch thicker than the disc. Attach the template to the face of the workpiece with screws, nails, or double-faced carpet tape. Adjust the height of the sanding drum about $\frac{1}{16}$ inch above the disc, turn on the drill press, and sand the perimeter of the workpiece. When the workpiece is the proper shape, the template will rub against the disc and prevent you from removing any more stock.

A SAFETY REMINDER

When you feed wood past a drum sander, always feed *against* the rotation of the drum. This is especially important when using a fence to sand to width or thickness. If you feed a workpiece between the fence and the sanding drum in the same direction that the drum is spinning, the machine will yank the wood out of your hands and launch it like a missile.

There are also soft-sided sanding drums with surfaces backed by flexible materials or air-filled bladders. These allow you to blend surfaces, round over sharp corners, and sand gently curved three-dimensional contours. Unlike the harder variety, these soft sanding drums are often used freehand. Instead of supporting the stock on a flat surface, hold it in your hands and press it against the abrasive. (*SEE FIGURE 5-6.*)

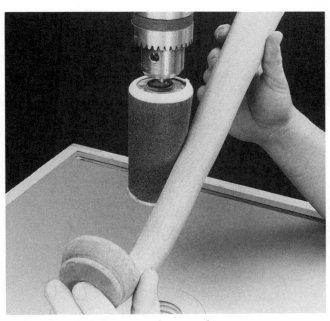

5-6 Soft-sided drum sanders, such as this pneumatic drum, allow you to sand gentle three-dimensional curves and contours. You can control the firmness of the abrasive surface by changing the air pressure inside the drum. The more air you pump in, the harder the drum will become. Let some air out, and the drum will become softer. The softer the surface, the easier it is for the drum to conform to curves.

ROUTING AND SHAPING

The drill press makes an acceptable stationary router — just mount a router bit in place of the drill bit. If you can, purchase a collet to hold the router bit. (Several tool manufacturers offer these as accessories for their drill presses.) If a collet isn't available, make your own from a standard bushing. (*SEE FIGURES 5-7 AND 5-8.*)

5-7 To safely mount router bits with ¼-inch-diameter shanks in a three-jaw drill chuck, you must make a simple collet. Purchase a bronze bushing with a ½-inch outside diameter and a ¼-inch inside diameter. Using a hacksaw, cut a slot in the bushing, parallel to the length. Remove any burrs on the inside of the bushing with a needle file. Insert the router bit's shank in the split bushing, then secure the bushing in the chuck. As the jaws squeeze the bushing, it will close around the shank like a router collet. The bushing not only locks the bit in place but also buttresses the slender shank to prevent it from bending or breaking. **Note:** Bits with ½-inch-diameter shanks are sturdy enough to be mounted directly in the chuck without a collet. Although the chuck does not support the shanks as well or grip them as tightly as a router collet, there is little danger they will break or slip at drill press speeds.

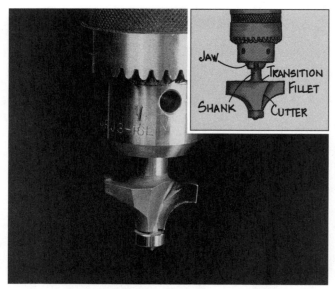

5-8 When mounting a router bit in a drill press chuck, you must be careful not to catch the *transition fillet* in the jaws. (This fillet is located between the bit's cutting edges and its shank.) This will prevent the bushing or the chuck jaws from getting a good grip on the shank, and the bit may slip while you're working. To avoid this, don't insert the bit's shank all the way into the chuck. Leave at least ⅛ inch of the shank showing. To further guard against slipping, tighten the chuck *twice,* inserting the key into two separate chuck holes and turning the ring gear.

5-9 When routing with unpiloted bits, use a fence to guide the work. *Never* pass the work between the bit and the fence — this exposes the entire bit. Instead, position the fence on the drill press table so most of the bit will be "buried" in the cutout. You will also need a cutout work surface should it be necessary to bury a part of the bit beneath the table. To further protect yourself, attach a guard to the quill to cover the exposed portion of the cutter. Adjust the vertical position of the guard so it just clears the work as you feed it past the bit. **Note:** The routing fence and cutout work surface shown are part of the "Drill Press Table" on page 96.

Depending on the bits you use and the routing operations you perform, you will need a work surface and a routing fence to fit your drill press table. Both the work surface and the fence must have a cutout so you can "bury the bit" — that is, position the unused portion of the cutter beneath the surface or behind the fence face. You must also make a guard to protect yourself from the exposed part of the cutter. (*See* Figure 5-9.)

When routing with the drill press, always feed the work against the rotation of the bit. Also remember that it runs much slower than standard routers, even at its highest setting. To make a clean, precise cut, you must adjust the feed rate to compensate. Feed the work much more slowly than you would when cutting with a router — this keeps the number of cuts per inch high and makes the cut smooth.

You also can get cleaner results by taking smaller bites. Make deep cuts in several passes, removing just ⅟₁₆ to ⅛ inch of stock with each pass — much less than you would cut away with a standard router. This reduces the side thrust generated by feeding the work into the bit horizontally, keeps vibration to a minimum, and maintains a precise, even cut.

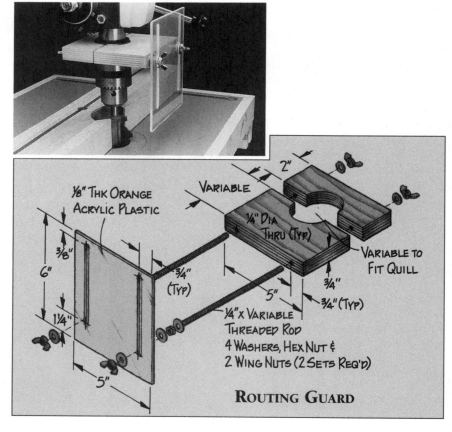

ROUTING GUARD

FOR YOUR INFORMATION

Several manufacturers make special routing bits with multiple flutes to run at drill press speeds. Unfortunately, only a limited number of shapes and sizes are available. You can purchase these from:

Wolfcraft, Inc.
P. O. Box 687
Itasca, IL 60143

By feeding the work slowly and removing small amounts of stock, you increase the time it takes to make an acceptable cut. Consequently, the drill press isn't an efficient routing tool — don't plan on retiring your router or router table. However, because this machine holds the bit *over* the work, it performs a few routing operations more easily than an ordinary table-mounted router.

For example, if you need to duplicate patterned workpieces, you can *pin rout* the patterns on the drill press, using a pin and a template to guide the work. Mount a straight bit in the chuck and secure a metal pin to the table directly below it. The bit and the pin should be the same diameter. Make a template for the pattern you want to rout, and attach the template to the workpiece. Then guide the template past the pin

as you rout the workpiece — the completed part will be the same shape as the template. (*SEE FIGURES 5-10 THROUGH 5-12.*) Use this technique to duplicate complex patterns, make signs, rout curved grooves for tambours, and cut recesses for inlays.

The drill press is also a good tool for routing mortises and slots in turnings, round stock, and odd-shaped parts. To rout a turning or a cylinder, secure it in a V-jig. For other shapes, fashion a special jig to hold the part. Feed the workpiece under the bit, guiding the jig along the fence. (*SEE FIGURE 5-13.*)

5-10 To pin rout patterns on the drill press, mount a router bit in the chuck. Adjust the table height so the distance between the bit and the table is the same as or a little less than the drill press stroke, then lock the table in place so it can't move. Attach a sheet of plywood or particleboard to the table and bore a stopped hole in it with the router bit. Cut a short length of metal rod that's the same size as the router bit to make a pin. Insert this pin into the hole — the pin will be perfectly centered under the bit.

5-11 Cut a pattern for the shape you want to rout, and fasten it to a sheet of plywood or hardboard. If you're making letters for a wooden sign, cut the first copy of each letter on a scroll saw and glue it to the sheet. Or, if you're restoring or repairing a piece of furniture, attach the part you want to replace to the sheet. All of these will serve as pin-routing templates. Attach the workpiece to the opposite side of the sheet with screws, nails, or double-faced carpet tape. **Note:** If you use metal fasteners, be careful not to place them where you might nick them with the router bit.

5-12 Adjust the depth stop to halt the bit when it's cutting $\frac{1}{16}$ to $\frac{1}{8}$ inch deep. Place the workpiece under the bit so the pin rests against the template. Turn on the drill press, feed the bit into the stock, and secure the quill lock. Rout the complete shape, letting the pin follow the template. Stop the machine, readjust the depth stop for a slightly deeper cut, and repeat. Continue until you have cut through the stock or to the depth needed.

5-13 To rout a mortise in a turned part, mount it in a V-jig. Use metal *hanger straps* (available in the plumbing departments of most hardware stores) to clamp the turning so it can't spin in the jig. Position the fence on the drill press to guide the work, and set the depth stop to halt the bit when it's cutting $\frac{1}{16}$ to $\frac{1}{8}$ inch deep. Clamp stops on the fence to limit the jig's lateral travel, allowing you to make repeat cuts easily. Place the turning under the bit, turn on the machine, and feed the bit into the wood. Guide the jig along the fence, cutting the mortise in the turning. Turn the drill press off, readjust the depth stop to cut a little deeper, and repeat. Continue until you have cut the mortise as deep as needed.

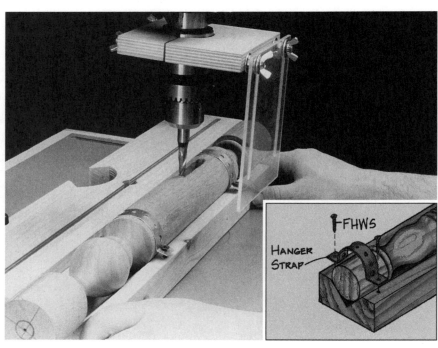

TRY THIS TRICK

Mount a ½-inch-diameter threaded arbor in the drill press and use the machine as a shaper. Several tool manufacturers sell these accessory arbors for mounting shaper cutters in table-mounted routers. Secure the arbor in the chuck instead, then mount a cutter on it. Use a cutout fence to guide the work.

TURNING

You can turn small wooden parts such as cabinet knobs and gallery spindles on a drill press, using the machine as a vertical lathe. To do this, you must make a drive center, a dead center, and a tool rest. Mount the drive center in the chuck, and clamp the dead center and the tool rest to the drill press table. Align the centers vertically. (*SEE FIGURE 5-14.*)

Secure the turning stock between the centers, using the quill to apply pressure. Turn on the drill press and rest a chisel against the tool rest. Slowly advance the

cutting edge of the chisel until you're cutting the wood. Make very light cuts, using the chisel to create the shapes you want. (*SEE FIGURES 5-15 AND 5-16.*)

5-14 To use your drill press as a lathe, make a drive center to turn the stock, a dead center to serve as a pivot, and a tool rest to support the chisel as you cut. Secure the dead center and the tool rest to a sheet of plywood. Mount the drive center in the chuck, and clamp the sheet to the drill press table so the dead center is directly beneath it. The points on the two parts should almost touch one another. **Note:** If you have a cutout table or a drill press table with a hole in the center, you can make an *adjustable* dead center, as shown. Turn the threaded rod to adjust the height of the dead center — raise it for short turnings; lower it for longer ones. Then lock it in place with a jam nut.

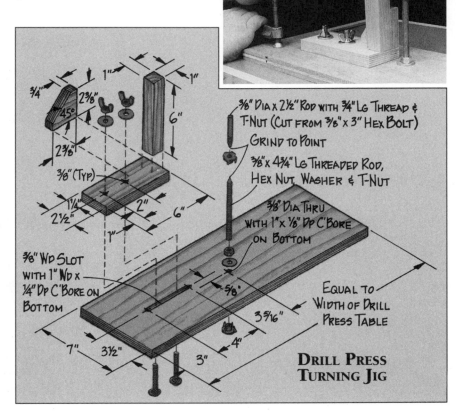

DRILL PRESS
TURNING JIG

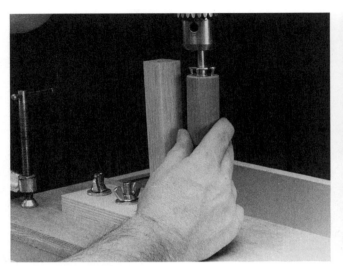

5-15 Cut the turning stock into short lengths, no longer than 12 inches (twice the length of the tool rest). Mark the center of the stock on the top and bottom ends and drill a $7/16$-inch-diameter, $1/2$-inch-deep hole in each end. Adjust the table height so the distance between the centers is slightly more than the length of the stock. Apply a little paste wax to the dead center, then place the stock so the hole in the bottom end fits over it. Advance the quill so the drive center engages the top hole, and apply enough pressure that the T-nut spurs bite into the wood. Secure the quill lock to keep the stock between the centers.

5-16 Position the tool rest as close to the stock as possible, but far enough away that the stock clears the rest by at least $1/8$ inch as it spins. Turn on the drill press, place a chisel against the rest, and slowly advance the tool until you're cutting the stock. Round the stock first, making a simple cylinder. Then turn the shapes. Use gouges to create coves (concave curves) and chisels to create beads (convex curves). **Note:** If the stock is longer than the tool rest, turn one portion, flip the stock end for end, and turn the other.

GRINDING AND SHARPENING

You can grind and hone cutting tools on a drill press with shank-mounted grindstones. These come in several different shapes — drums, wheels, cups, and cones — allowing you to grind different profiles and sharpen a variety of tools. For freehand grinding, make a rest and clamp it to the drill press table to help support and control the work. (SEE FIGURE 5-17.) When grinding flat surfaces and straight edges, position the fence to guide the work past the stone. (SEE FIGURES 5-18 THROUGH 5-20.) If necessary, make a fixture or a carriage to hold the work as you guide it along the fence. Tilt the table to grind or sharpen at an angle. (SEE FIGURES 5-21 AND 5-22.)

5-17 When grinding freehand, support the work with a rest, as shown. Place the work against the rest and use the rest as both a pivot and a backstop as you maneuver the work against the stone. **Note:** The lathe tool rest from the turning jig shown in FIGURE 5-14 on page 89 makes an excellent rest for grinding.

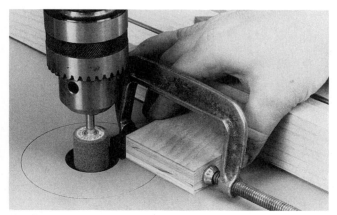

5-18 To grind a flat surface or
a straight edge, use a fence to guide
the work. If necessary, make a carriage to hold the work, then guide
the carriage along the fence and past
the grindstone. For example, to grind
a flat surface on a band-saw guide
block, clamp it to a block of wood.
Guide the wood block along a fence
past a grinding drum. Position the
fence so the wheel removes just a
tiny amount of metal from the face
of the guide block as it passes.

Set the drill press to run at a slow speed, especially if
you're sharpening a delicate tool. This will prevent the
stone from overheating the cutting edge and reducing
the temper. Set the fence or the cross vise to remove a
very small amount of metal with each pass — no more
than the thickness of a piece of paper. (This also keeps
the edge cool.) Feed the work at a slow, steady rate,
keeping it in motion at all times. Make several passes
before you change the position of the fence·or the vise
to remove more stock. As you work, lubricate the
stone with 10W oil to prevent the metal filings from
clogging the abrasive.

Try This Trick

When guiding work along a fence, use strips
of paper as shims to feed the work into the grindstone. After making the first pass, place a strip
between the fence and the work to position it 2 to
3 thousandths of an inch closer to the stone. Add
additional strips as needed.

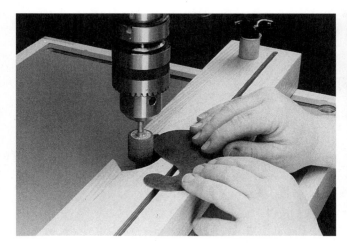

5-19 In some cases, you can use
the top of a fence as a rest to support
the work as you grind it. For example,
if you have a fence with a broad, flat
top, it makes short work of sharpening scrapers with curved edges.
Mount a grinding drum in the drill
press and secure the fence next to the
drum. (Or, if the fence has a cutout,
position the drum inside the cutout.)
Rest the scraper on top of the fence
and feed the edges past the drum,
grinding away the old, dull burrs.

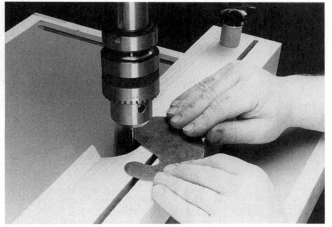

5-20 To raise new burrs on the
scraper, replace the grinding drum
with a hardened steel dowel pin
(available at most hardware stores).
Tilt the table 5 degrees, spread a drop
of oil on the dowel pin, and turn on
the drill press. Feed the scraper past
the lubricated pin, pressing the edges
firmly against it — this will raise a
burr, just like a burnisher. Raise a
burr all around the circumference,
then turn the scraper over and raise
a second burr on the other side.

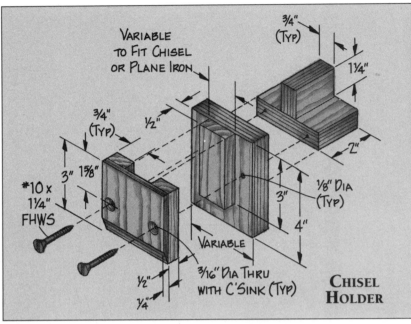

5-21 You can also sharpen chisels and plane irons with a grinding drum, but you must make a holder to support the tool. Clamp the tool in the holder, rest it against the fence, and tilt the table to the desired angle. (Chisels and irons are typically ground so the tool angle is between 25 and 30 degrees.) Position the fence to remove a tiny amount of metal, turn on the drill press, and feed the tool and holder past the drum.

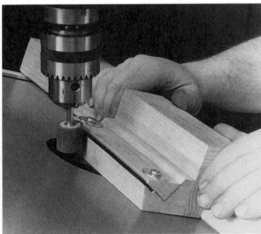

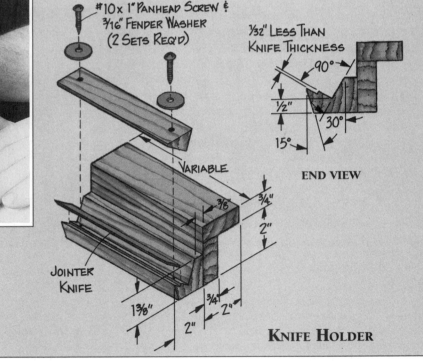

5-22 You can use a similar setup to sharpen jointer knives. Make a sharpening carriage by cutting a wedge from a block of wood. Insert a knife under the wedge and secure it with screws. Tilt the table to the desired angle — jointer knives are usually ground at 45 degrees. In the setup shown, the table is tilted to 15 degrees and the slot in the wood block is angled at 30 degrees, which adds up to 45 degrees. Position the fence to remove a tiny amount of metal, turn on the drill press, and feed the knife and holder past the drum. **Note:** Grind *all three* jointer knives without moving the fence. Then reposition the fence to remove more metal, if necessary, and grind all three again. This will ensure all three knives are ground identically.

QUILL LOCKS AND SIDE THRUST

There is much controversy among craftsmen concerning whether or not you can rout and shape on the drill press without harming the machine. According to the opponents, routing and shaping operations produce side thrust that quickly wears out the bearings and other parts of the press. Actually, *all* the operations in this chapter — routing, shaping, sanding, turning, grinding, and sharpening — produce side thrust, and that thrust *may* be detrimental if your machine isn't built to take it.

By pushing against the end of the extended quill, side thrust may cause *racking*. This, in turn, forces the moving parts out of alignment and causes excessive wear. For example, if there is only one bearing in the tip of the quill, the spindle will be racked to one side. Because they are forced slightly out of alignment, the moving parts will wear faster than they would otherwise. A two-bearing quill prevents the spindle from racking and allows no more wear than normal.

Side thrust can also cause the quill itself to rack inside the head casting, wearing the inside surfaces of the sleeve that guides it. As the opening becomes larger, the quill will no longer track accurately. You will find it progressively more difficult to position holes. This can be prevented with a quill lock that squeezes the quill inside the sleeve, keeping it from moving vertically *and* horizontally. There is no racking — and no wear on the inside surfaces of the sleeve.

In short, if your drill press has the equipment to limit the racking caused by side thrust — a two-bearing quill and a quill lock — you can safely rout, shape, or perform any of the other operations in this chapter.

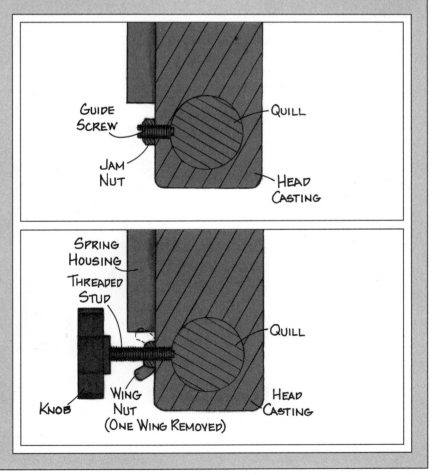

1 Most well-made drill presses have two-bearing quills, but not all of them have a quill lock. If yours lacks a lock, you can easily make one. Locate the guide screw in the side of the head casting. This screw rides in the slot in the side of the quill and keeps it from turning in the sleeve. It's typically locked in place with a jam nut. Remove the screw and replace it with a same-size thumbscrew. You can also use a knob and threaded stud. Substitute a wing nut for the jam nut to keep the thumbscrew (or the stud) from vibrating loose during normal drilling operations. **Note:** If the hole for the guide screw is too close to the return spring housing (as shown here), you may have to cut one of the wings off the wing nut.

(continued) ▷

QUILL LOCKS AND SIDE THRUST — CONTINUED

2 **To make this substitution,** you must create a nub on the end of the thumbscrew to fit the vertical slot in the quill. The diameter of this nub should equal the width of the slot; and the nub's length should equal the depth of the slot. Use two nuts jammed together as a stop to set the length of the nub, then carefully file away the threads showing above the nut.

3 **To use the quill lock, simply** tighten the thumbscrew or turn the knob. The nub will press the quill sideways against the inside surface of the sleeve, preventing it from moving. For normal drilling operations, loosen the quill lock a fraction of a turn so the quill feeds and retracts smoothly. Tighten the wing nut against the head casting to jam the quill lock in place.

PROJECTS

6

DRILL PRESS TABLE

Most experienced woodworkers agree that the standard drill press table has serious shortcomings. It's too small to support large and medium-size workpieces properly. There's no easy way to attach a fence, so it's difficult to hold and guide a workpiece. And it tilts in the wrong direction — side to side instead of front to back. This makes it difficult to drill angled holes in long workpieces.

The shop-made drill press table shown bolts to a standard table and makes a marked improvement:

■ It offers a generous work surface, 24 inches wide and 18 inches deep.

■ The table is slotted so you can easily mount and position a fence.

■ The table tilts front to back, up to 45 degrees from horizontal.

And there are several additional user-friendly features:

■ The fence-mounting system is almost instant — just insert the heads of the carriage bolts in the access holes at the front ends of the slots. And there are mounts under the table to store the fence and keep it handy when it's not in use.

■ You can also use the table slots to mount clamps and hold-downs.

■ The top of the fence is slotted to hold stops and other shop-built drilling fixtures.

■ The fence is reversible and has a cutout in the back so you can use it for routing and shaping.

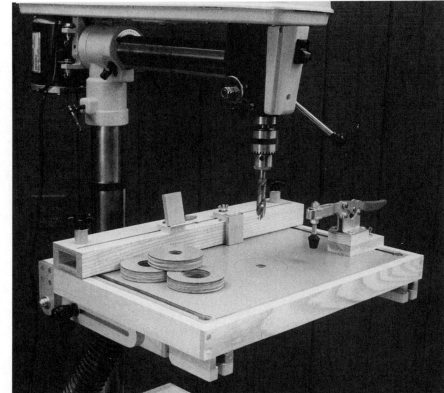

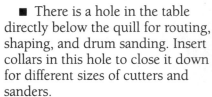

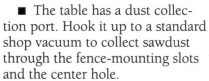

■ There is a hole in the table directly below the quill for routing, shaping, and drum sanding. Insert collars in this hole to close it down for different sizes of cutters and sanders.

■ The table has a dust collection port. Hook it up to a standard shop vacuum to collect sawdust through the fence-mounting slots and the center hole.

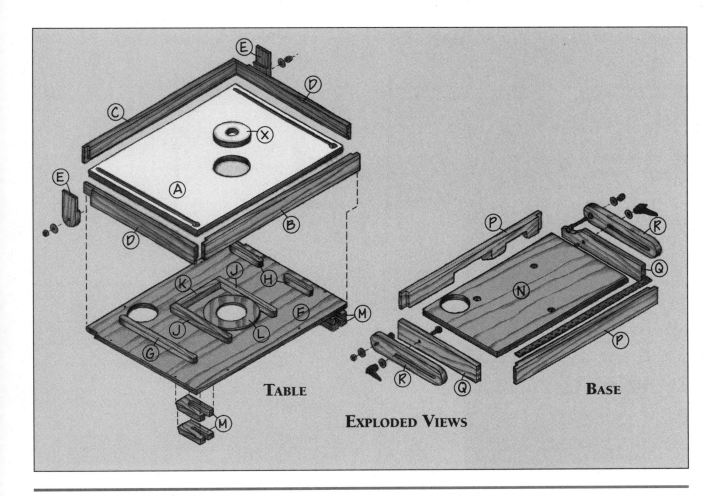

TABLE

BASE

EXPLODED VIEWS

MATERIALS LIST (FINISHED DIMENSIONS)

Parts

Table

A. Tabletop*   ¾" x 16½" x 22½"
B. Table front   ¾" x 2" x 24"
C. Table back   ¾" x 2" x 23¼"
D. Table sides (2) ¾" x 2" x 17¼"
E. Tilt arm
    mounts (2)   ¾" x 2" x 4"
F. Bottom*   ½" x 17¼" x 23¼"
G. Long baffle   ¾" x ¾" x 12½"
H. Short baffles (2) ¾" x ¾" x 4"
J. Center hole
    side baffles (2) ¾" x ¾" x 8¼"
K. Center hole
    back baffle   ¾" x ¾" x 6"
L. Center hole
    collar support*   6" dia. x ¼"
M. Fence mount tops/
    bottoms (4)   ¾" x 2" x 3¾"

Base

N. Base*   ¾" x 11¼" x 19½"
P. Base
    front/back (2)   ¾" x 2" x 21"
Q. Base sides (2)   ¾" x 2" x 12"
R. Tilt arms (2)   ¾" x 2" x 10¾"

* Make these parts from plywood.

Hardware

Table

#8 x 1¼" Flathead wood screws
    (16)
#8 x ¾" Flathead wood screws
    (12)
#6 x 1" Flathead wood screws (10)
18" x 24" Plastic laminate
    (optional)

Base

⅜" x 2" Carriage bolts (3)
5/16" x 2" Carriage bolts (2)
5/16" x 2" Hex bolts (2)
⅜" Flat washers (3)
⅜" Fender washers (3)
5/16" Flat washers (4)
⅜" Wing nuts (3)
5/16" Stop nuts (2)
5/16" Ratchet handles (2)
1½" x 20" Piano hinge and
    mounting screws

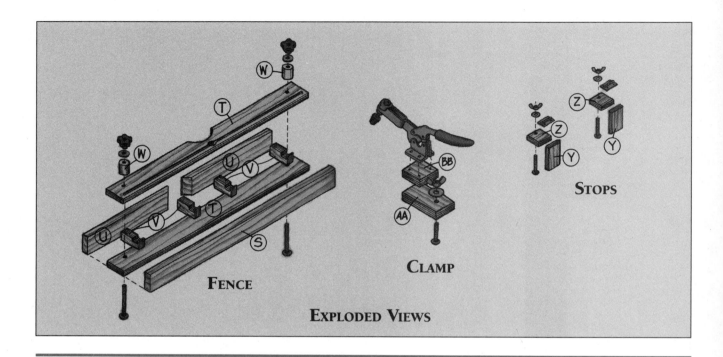

FENCE

CLAMP

STOPS

EXPLODED VIEWS

MATERIALS LIST (FINISHED DIMENSIONS)

Parts

Fence

S. Fence solid face   ¾″ x 2″ x 24″
T. Fence top/
   bottom*   ½″ x 2¼″ x 24″
U. Fence split
   faces (2)   ¾″ x 2″ x 11″
V. Fence
   braces (4)   ¾″ x 1″ x 2¼″
W. Locking knob
   spacers (2)   1″ dia. x 1″

Accessories

X. Collars* (4)   4″ dia. x ¾″
Y. Stops (2)   ½″ x 1⅝″ x 2⅜″

Z. Stop
    mounts (2)   ½″ x 1½″ x 1⅝″
AA. Clamp base   ¾″ x 2¼″ x 4½″
BB. Clamp mount   ¾″ x 2¼″ x 2¼″

* *Make these parts from plywood.*

Hardware

Fence

⅜″ x 4½″ Carriage bolts (2)
⅜″ Flat washers (2)
⅜″ Star knobs (2)

Accessories

Stops (2)
¼″ x 1½″ Carriage bolts (2)
¼″ Flat washers (2)
¼″ Wing nuts (2)
1″ x 1½″ Butt hinges and
   mounting screws (2)

Clamp
⅜″ x 2″ Carriage bolt
⅜″ Flat washer
⅜″ Wing nut
Toggle clamp and mounting screws

PLAN OF PROCEDURE

1 Select the materials and cut the parts to size.
To make the drill press table, fence, and accessories, you need about 5 board feet of 4/4 (four-quarters) hardwood lumber, one-quarter sheet of ¾-inch plywood, one-quarter sheet of ½-inch plywood, a scrap of ¼-inch plywood, and a scrap of 1-inch-diameter wooden dowel rod. The table shown is made from ash

lumber and birch plywood, but you can use any *cabinet-grade* plywood and hardwood.

If you wish, cover the tabletop with plastic laminate to make it more durable. Plane the lumber to ¾ inch thick, and cut all parts to the sizes given in the Materials List. Carefully label each part to keep from mixing them up.

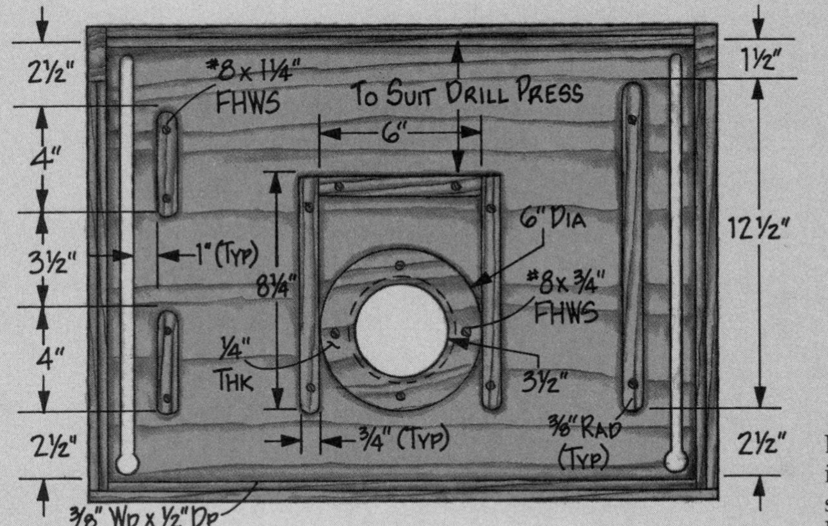
2½"
4"
3½"
4"
2½"
#8 x 1¼" FHWS
To Suit Drill Press
6"
1" (Typ)
8¼"
¼" Thk
6" Dia
#8 x ¾" FHWS
3½"
⅜" Rad (Typ)
¾" (Typ)
⅜" Wd x ½" Dp Rabbet (Typ)
1½"
12½"
2½"

NOTE: Bottom is removed to show baffle layout.

BOTTOM VIEW
TABLE

MAKING THE TOP

2 Cut the joinery in the table parts. The tabletop, bottom, front, back, sides, and tilt arm mounts are all assembled with simple rabbets. Make these joints with a table-mounted router or a dado cutter:

- 2-inch-wide, ⅜-inch-deep rabbets in the ends of the tilt arm mounts, as shown in the *Tilt Arm Mount Layout*
- 1⅝-inch-wide, ⅜-inch-deep rabbets in the ends of the table sides to join the tilt arm mounts
- ¾-inch-wide, ⅜-inch-deep rabbets in the ends of the front to fit the sides, as shown in the *Table/Top View*
- ⅜-inch-wide, ½-inch-deep rabbets in the bottom edges of the front, back, and sides to hold the bottom, as shown in the *Table/Bottom View*
- ⅜-inch-wide, ⅜-inch-deep rabbets in the ends of the back to fit the sides and the tilt arm mount

3 Drill or cut holes in the tabletop, bottom, tilt arm mount, and collar support. You must make several holes in the table parts for dust collection or hardware:

- a 2¼-inch-diameter hole in the table bottom to hook up a vacuum hose, as shown in the *Table Bottom Layout*
- ⅞-inch-diameter holes in the tabletop to mark the front ends of the fence-mounting slots, as shown in the *Table/Top View*
- ⅜-inch-diameter holes in the tabletop to mark the back ends of the fence-mounting slots
- 5⁄16-inch-diameter holes in the ends of the tilt arm mounts, as shown in the *Tilt Arm Mount Layout*
- ⅛-inch-diameter pilot holes with countersinks in the table bottom

Do not cut the 4-inch-diameter center hole in the tabletop yet; wait until after you have mounted the table on the drill press.

4 Cut the slots in the tabletop and fence mounts. The fence mounts in ⅜-inch-wide slots in the tabletop. These slots end in the ⅞-inch-diameter holes at the front of the table, letting you insert the heads of the fence-mounting bolts. The slots also help collect sawdust as you work. Make the slots with a router and a ⅜-inch straight bit. Using a band saw, also cut:

- ¾-inch-wide slots in the fence mount tops, as shown in the *Fence Mount Detail*
- ⅜-inch-wide slots in the fence mount bottoms

5 Cut the profiles of the collar support, tilt arm mounts, fence mounts, and baffles. Glue the fence mount tops to the fence mount bottoms, and lay out the bevel on the back ends. Also lay out the circular shape of the collar support and the rounded ends of the tilt arm mounts. Cut the bevels and the profiles with a band saw or a saber saw, then sand the sawed

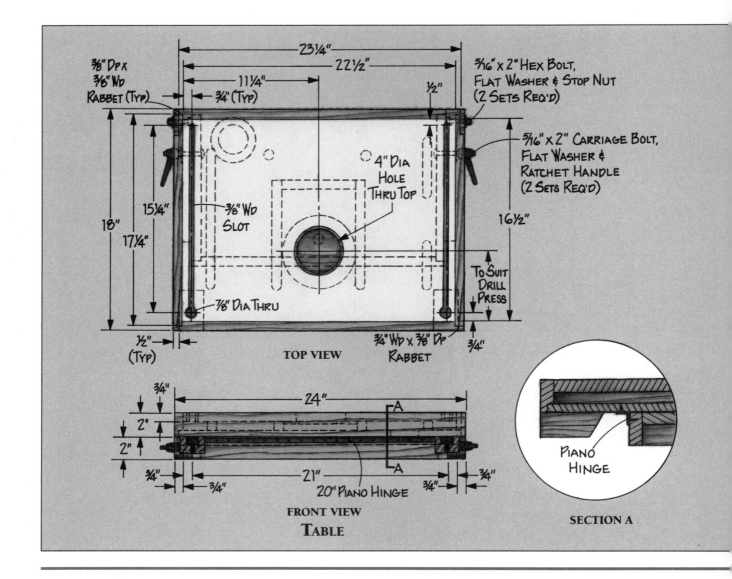

TOP VIEW

FRONT VIEW
TABLE

PIANO HINGE

SECTION A

surfaces. While you're at the sander, round over both ends of the long baffle and short baffles, and one end of each center hole side baffle, as shown in the *Table/Bottom View*.

6 Assemble the table. Finish sand the wooden parts of the table. Glue together the top, front, back, sides, and tilt arm mounts. When the glue dries, sand all joints clean and flush.

Screw the fence mounts to the bottom, making sure that the distance between the ⅜-inch-wide slots in the mounts is exactly the same as the distance between the slots in the work surface. Attach the bottom to the table assembly with #6 flathead wood screws. Do *not* glue the bottom in place; you may have to remove it periodically to clean the inside of the table.

MAKING THE BASE

7 Cut the joinery in the front and the back. Cut ¾-inch-wide, ⅜-inch-deep rabbets in the ends of the base front and back, using a table-mounted router or a dado cutter. These rabbets join the sides to the front and back.

8 Drill the holes in the base parts. Bore or cut the holes needed in the base and the base sides:

■ a 3½-inch-diameter hole in the base top, as shown in the *Base/Bottom View*, to let you attach a vacuum hose directly to the table bottom

■ ⅜-inch-diameter holes with 1-inch-diameter counterbores in the base to fasten the base to the drill press table

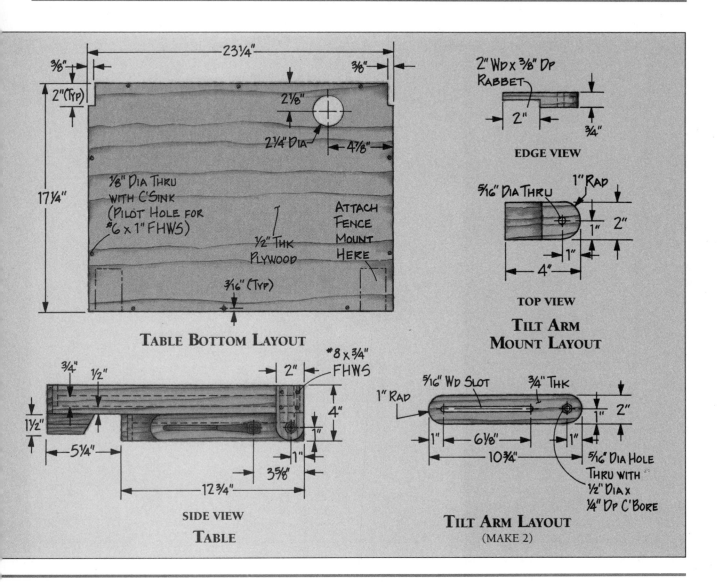

TABLE BOTTOM LAYOUT

EDGE VIEW

TOP VIEW

TILT ARM MOUNT LAYOUT

SIDE VIEW
TABLE

TILT ARM LAYOUT
(MAKE 2)

■ ⁵⁄₁₆-inch-diameter holes with ½-inch-diameter counterbores in the tilt arms, as shown in the *Tilt Arm Layout*

■ ⁵⁄₁₆-inch-diameter holes to mark the ends of the slots in the tilt arms

■ ⁵⁄₁₆-inch-diameter holes in the base sides to secure the tilt arms

9 Cut slots in the tilting arms. As you change the angle of the table, the tilting arms slide up and down on ⁵⁄₁₆-inch-wide slots, as shown in the *Tilt Arm Layout*. Cut these slots with a router and a ⁵⁄₁₆-inch straight bit.

10 Cut the base back to fit around the drill press table. Depending on the size and shape of your

drill press table and its supporting arm, you may have to cut a notch in the base back to fit around it. And you may have to make another notch to accommodate the table height adjustor crank. Dry assemble (without glue) the base, front, back, and sides, and hold the parts together with a band clamp. Lay the base assembly in place on the drill press table to see if it will fit as is. If not, make the necessary cutouts in the back with a band saw or a saber saw. Start small, then enlarge them until you get the fit you want. Make sure the notch for the crank is large enough that you won't bang your knuckles when you adjust the table height.

11 Assemble the base and attach the table. Finish sand the wooden parts, and glue the base,

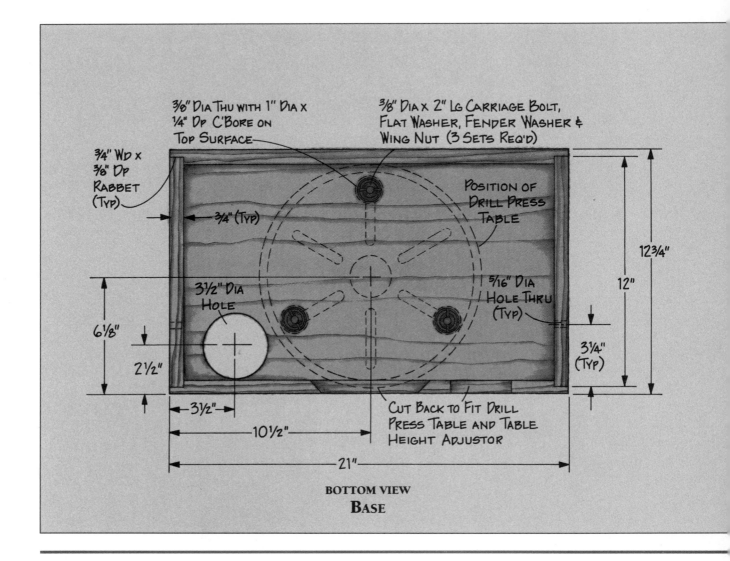

3/8" Dia Thu with 1" Dia x 1/4" Dp C'Bore on Top Surface

3/8" Dia x 2" Lg Carriage Bolt, Flat Washer, Fender Washer & Wing Nut (3 Sets Req'd)

3/4" Wd x 3/8" Dp Rabbet (Typ)

3/4" (Typ)

Position of Drill Press Table

3 1/2" Dia Hole

5/16" Dia Hole Thru (Typ)

12 3/4"

12"

6 1/8"

2 1/2"

3 1/4" (Typ)

Cut Back to Fit Drill Press Table and Table Height Adjustor

3 1/2"

10 1/2"

21"

BOTTOM VIEW
BASE

front, back, and sides together. When the glue dries, sand the parts clean and flush.

Attach the table to the base with a piano hinge, as shown in the *Table/Section A*. (When the table is horizontal, the back of the table should be flush with the back of the base.) Insert hex bolts in the tilt arms, and use a nut and a flat washer to draw each hex head into its counterbore. Secure the tilt arms to their mounts with washers and stop nuts, and to the base sides with carriage bolts, washers, and ratchet handles.

Mount the table and base to the drill press table. Test the tilting action of the table. Loosen the ratchet handles and lift up on the back of the table. (If the movement seems stiff, the stop nuts that hold the tilt arms to the mounts may be too tight.) When the table is tilted as far as it will go, it should be just a little past 45 degrees off horizontal.

FOR YOUR INFORMATION

You can set the table to 45 degrees quickly by using a small square to set the tilt arms perpendicular to the base.

12 **Cut the center hole in the table.** With the table mounted to the drill press and centered under the quill, mount a bit with a lead point in the chuck. Extend the quill until the lead point touches the table — this indicates the center of the large hole in the table.

Dismount the table from the drill press. Remove the table assembly from the base assembly, and strip all the hardware — bolts, washers, nuts, handles, and hinges. Set the hardware aside.

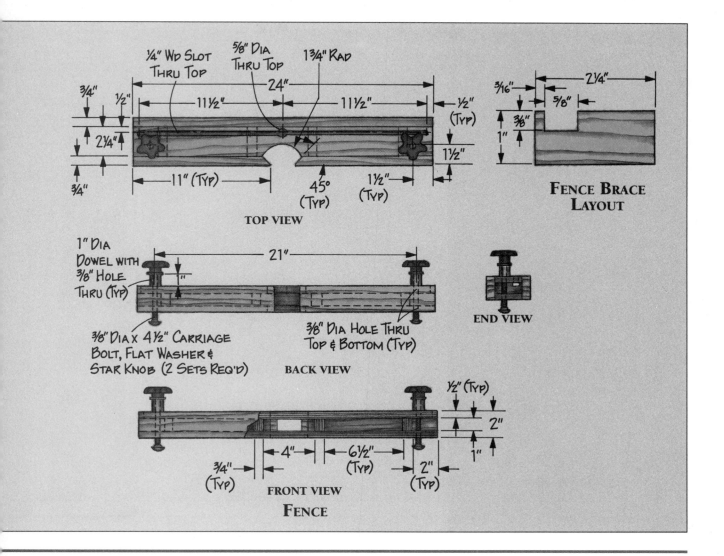

Cut a 4-inch-diameter hole in the table where the lead point touched the surface. Attach the collar support and the baffles to the underside of the tabletop with glue and #8 flathead wood screws. (These baffles direct the air flow when the table is hooked to a dust collector, keeping the suction even along the length of the slots and at the center hole.)

Don't reassemble the table and the base yet. Leave them in pieces until after you have applied a finish.

MAKING THE FENCE

13 **Drill the holes in the fence top, bottom, and spacers.** Lay the fence bottom across the tabletop, using a square to position it perpendicular to the sides. With a pencil, reach up through the slots and mark the bottom where it crosses the slots' sides.

Stack the top on the bottom, holding the two parts together with double-faced carpet tape. Lay out the ⅜-inch-diameter holes on the bottom, positioning them between the marks you just made. Drill the holes through both the top and the bottom, then take the parts apart and discard the tape. Also drill:

■ ⅜-inch-diameter holes in the locking knob spacers, as shown in the *Fence/Back View*

■ a ⅝-inch-diameter hole in the fence top, as shown in the *Fence/Top View*, to help mount stops and other accessories

■ ¼-inch-diameter holes to mark the ends of the slot in the fence top

14 **Cut the slots, notches, cutouts, and bevels in the fence parts.** Rout a ¼-inch-wide slot in the fence top, using a router and a straight bit. With a

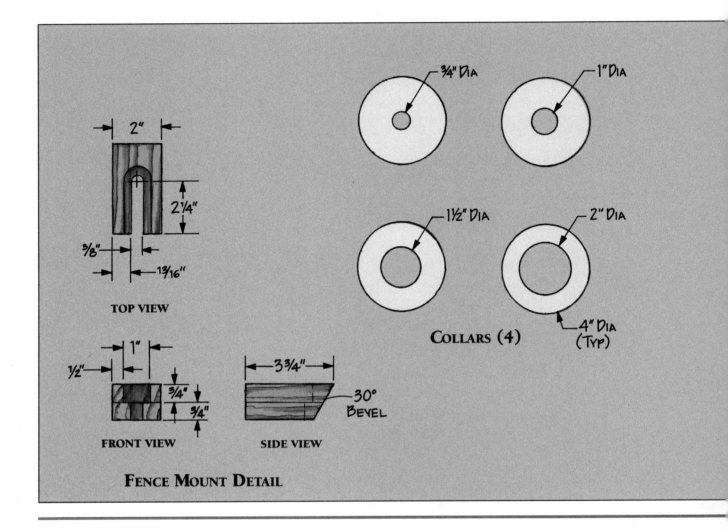

TOP VIEW

2"

2¼"

⅜"

1³⁄₁₆"

¾" DIA

1" DIA

1½" DIA

2" DIA

4" DIA
(TYP)

COLLARS (4)

½"

1"

¾"

¾"

3¾"

30°
BEVEL

FRONT VIEW

SIDE VIEW

FENCE MOUNT DETAIL

band saw or a saber saw, make the notches in the fence braces, as shown in the *Fence Brace Layout,* and the semicircular cutouts in the fence top and bottom, as shown in the *Fence/Top View.* On a table saw, bevel the inside ends of the split faces.

15 **Assemble the fence.** Finish sand the wooden parts of the fence, then assemble the top, bottom, solid face, and split faces with glue. Let the glue dry, and sand the joints clean and flush. True the solid face and the split faces on a jointer so they are perfectly straight and square to the bottom. Insert carriage bolts through the fence and the spacers, and secure the bolts with star knobs and flat washers.

Mount the fence to the table by inserting the heads of the carriage bolts in the holes at the table front and then sliding the fence back along the slots. Lock the table in position by tightening the knobs. To reverse the fence, dismount it and turn it face for face.

MAKING THE ACCESSORIES

16 **Make the collars.** If you have covered the table surface with plastic laminate, cover the collar stock, too — the table and the collars must be precisely the same thickness. Cut several collar blanks and stack them face to face, holding the stack together with double-faced carpet tape. Lay out the circular shape on the top blank and cut the entire stack with a band saw. Saw a little wide of the line, then sand the sawed edge up to the line. Take the stack apart, discard the tape, and drill a different-size hole in the center of each collar to accommodate various diameters of drum sanders and router bits.

17 **Make the stops.** Chamfer one edge of each stop — this will help prevent sawdust from interfering with the accuracy of the stop. Drill ¼-inch-diameter holes in the stop mounts, then attach the

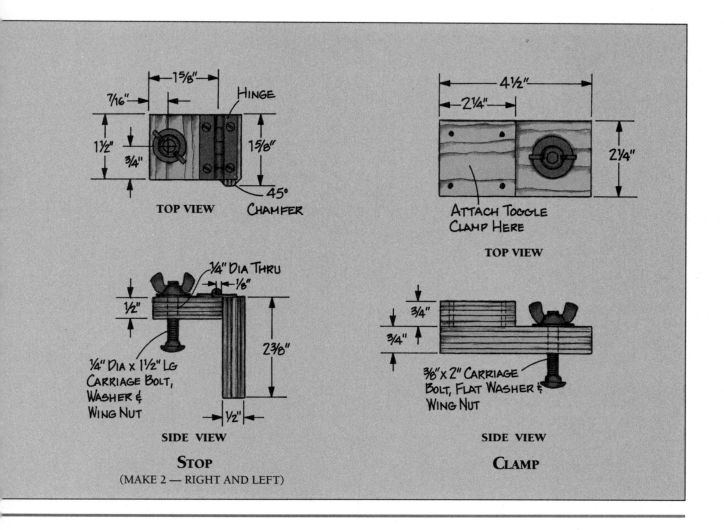

TOP VIEW

TOP VIEW

STOP
(MAKE 2 — RIGHT AND LEFT)

CLAMP

stops to the mounts with butt hinges. (The hinges let you flip the stops out of the way without having to remove them from the fence.) Mount the assembled stops in the slot in the top of the fence with carriage bolts, washers, and wing nuts.

18 Make a clamp. Drill a ⅜-inch-diameter hole in the clamp base and glue it to the clamp mount. Attach a toggle clamp to the mount. Secure the assembled clamp in one of the table slots with a carriage bolt, washer, and wing nut.

FINISHING UP

19 Apply a finish to the table, base, fence, and accessories. Disassemble the fence and the other accessories, removing the hardware and setting it aside. Do any necessary touch-up sanding to the wooden surfaces, then apply several coats of finish. When the finish dries, rub it out with steel wool and paste wax. Reassemble the drill press table, base, fence, and accessories. Fasten the assembled table and base to the drill press table.

TRY THIS TRICK

The wooden parts of the table shown are finished with a mixture of tung oil and spar varnish. Combine the two finishes in a ratio of 1 cup of tung oil to 1 tablespoon of spar varnish. Wipe or brush the finish on. Allow it to soak into the wood for several minutes, then wipe any excess off. Wait overnight, and apply a second coat. This makes an extremely durable finish for jigs and fixtures.

VARIATIONS

If you don't need a drill press table with all these whistles and bells, you can make a much simpler one. Cut a sheet of plywood 24 inches long and 18 inches wide — the same size as the other table. Rout two slots along the sides in the same position as the slots in the other design, then bolt the plywood directly to the drill press table.

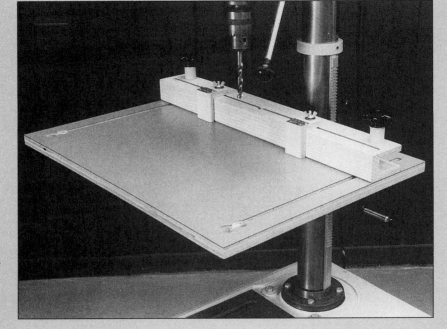

1 This simplified table will mount the same fence, stops, and clamp as the high-end version. When you build the fence, however, don't bother with the split faces or the cutouts for the router bits; you won't need them.

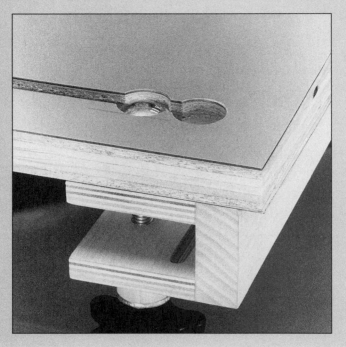

2 This design even includes a place to store the fence when it's not in use. Mount it to the underside of the table so the heads of the carriage bolts rest in the counterbores, as shown.

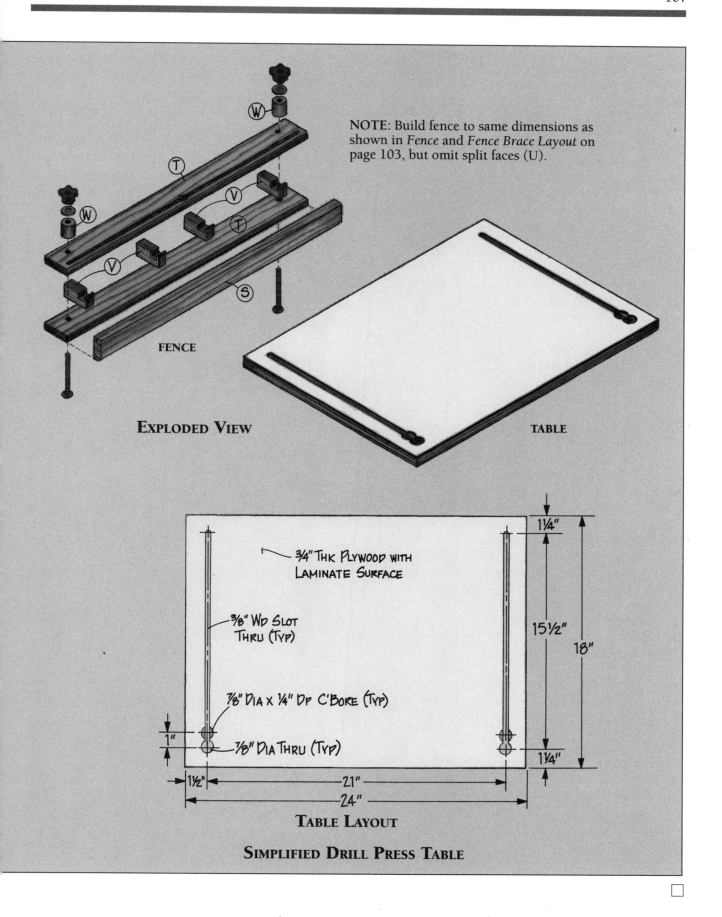

NOTE: Build fence to same dimensions as shown in *Fence* and *Fence Brace Layout* on page 103, but omit split faces (U).

FENCE

EXPLODED VIEW

TABLE

3/4" THK PLYWOOD WITH LAMINATE SURFACE

3/8" WD SLOT THRU (TYP)

7/8" DIA x 1/4" DP C'BORE (TYP)

7/8" DIA THRU (TYP)

1 1/4"

15 1/2"

18"

1 1/4"

1"

1 1/2"

21"

24"

TABLE LAYOUT

SIMPLIFIED DRILL PRESS TABLE

7

DRILL PRESS STAND

There may be more accessories available for the drill press than for any other power tool, and without a way to store them, you can find yourself chasing drill bits and cutters all over your shop.

This drill press stand solves your storage problems simply and effectively. It's just a plywood box, sized to hold a bench drill press or a radial drill press at a comfortable working height. Inside the box, there are six drawers of various depths. This arrangement offers enough storage space to hold *all* of the accessories we've shown in this book — and still have drawer space left over!

Furthermore, the stand is easy and economical to build. You don't need to purchase expensive drawer-mounting hardware; the drawers slide in and out of ordinary grooves. And there are no fancy joints to make; the box and the drawers are assembled with simple rabbets and dadoes.

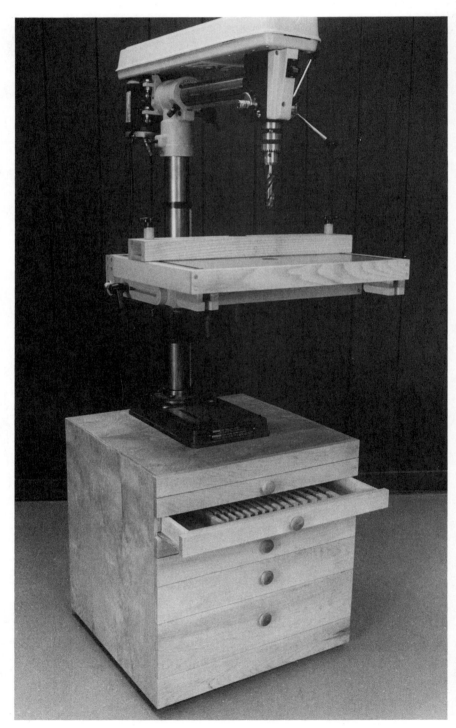

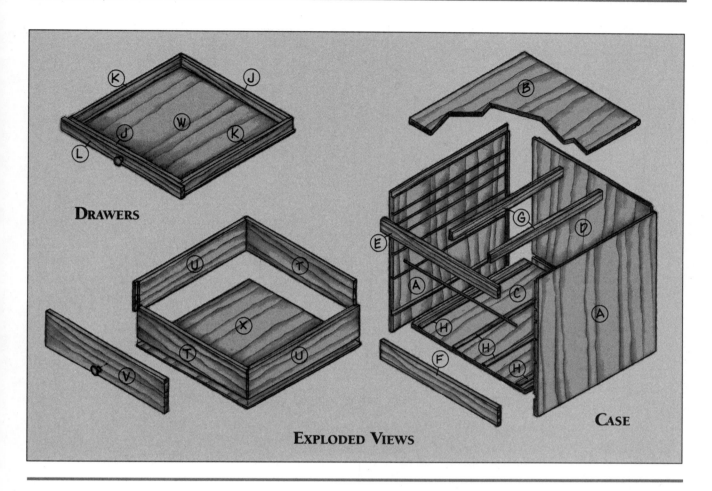

DRAWERS

EXPLODED VIEWS

CASE

MATERIALS LIST (FINISHED DIMENSIONS)

Parts

Case

A. Sides* (2) $\frac{3}{4}"$ x $23\frac{1}{4}"$ x $24"$
B. Top* $\frac{3}{4}"$ x $23\frac{1}{4}"$ x $23\frac{1}{4}"$
C. Bottom* $\frac{3}{4}"$ x $22\frac{7}{8}"$ x $23\frac{1}{4}"$
D. Back* $\frac{3}{4}"$ x $23\frac{1}{4}"$ x $23\frac{5}{8}"$
E. Valance $\frac{3}{4}"$ x $2"$ x $24"$
F. Kickboard $\frac{3}{4}"$ x $3\frac{1}{8}"$ x $24"$
G. Braces (2) $\frac{3}{4}"$ x $1\frac{1}{4}"$ x $22\frac{1}{2}"$
H. Drawer
   runners (4) $\frac{1}{8}"$ x $\frac{3}{4}"$ x $22\frac{1}{2}"$

Drawers

J. Top drawer fronts/
   backs (6) $\frac{1}{2}"$ x $1\frac{3}{4}"$ x $21\frac{7}{8}"$
K. Top drawer
   sides (6) $\frac{1}{2}"$ x $1\frac{3}{4}"$ x $22\frac{3}{8}"$
L. Top drawer
   faces (3) $\frac{3}{4}"$ x $2"$ x $24"$

M. Middle top drawer front/
   back (2) $\frac{1}{2}"$ x $2\frac{3}{4}"$ x $21\frac{7}{8}"$
N. Middle top drawer
   sides (2) $\frac{1}{2}"$ x $2\frac{3}{4}"$ x $22\frac{3}{8}"$
P. Middle top
   drawer face $\frac{3}{4}"$ x $3"$ x $24"$
Q. Middle bottom drawer front/
   back (2) $\frac{1}{2}"$ x $3\frac{1}{4}"$ x $21\frac{7}{8}"$
R. Middle bottom drawer
   sides (2) $\frac{1}{2}"$ x $3\frac{1}{4}"$ x $22\frac{3}{8}"$
S. Middle bottom drawer
   face $\frac{3}{4}"$ x $4"$ x $24"$
T. Bottom drawer front/
   back (2) $\frac{1}{2}"$ x $5\frac{1}{8}"$ x $21\frac{7}{8}"$
U. Bottom drawer
   sides (2) $\frac{1}{2}"$ x $5\frac{1}{8}"$ x $22\frac{3}{8}"$
V. Bottom drawer
   face $\frac{3}{4}"$ x $5\frac{7}{8}"$ x $24"$

W. Top, middle top, and middle
   bottom drawer bottoms*
   (5) $\frac{1}{4}"$ x $23\frac{1}{8}"$ x $22\frac{3}{8}"$
X. Bottom drawer
   bottom* $\frac{1}{4}"$ x $22\frac{3}{8}"$ x $22\frac{3}{8}"$

* Make these parts from plywood.

Hardware

#8 x $1\frac{1}{4}"$ Flathead wood screws
   (16–20)

4d Finishing nails (48–54)

$1\frac{1}{2}"$ Drawer pulls and mounting
   screws (6)

$\frac{3}{8}"$ dia. x $23"$ Steel rod

$3"$ Swiveling casters and mounting
   screws (4)

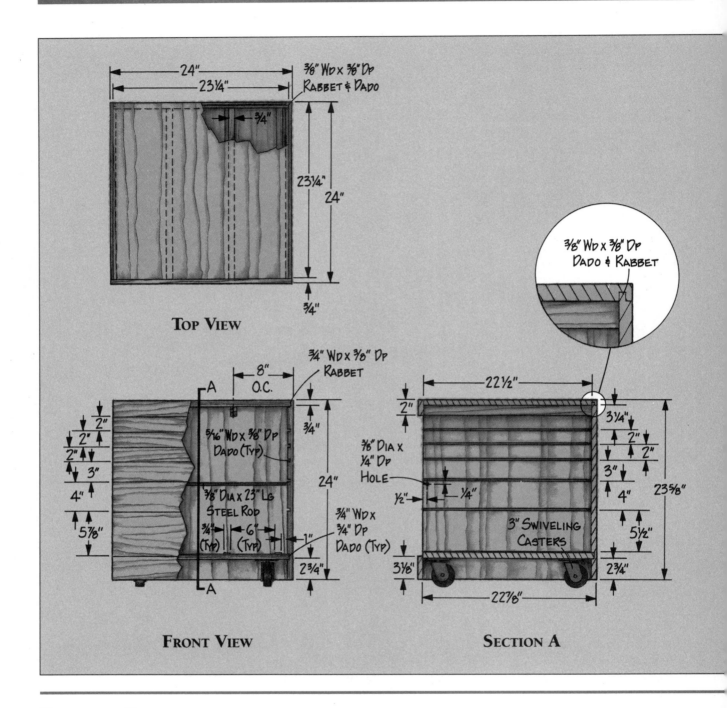

TOP VIEW

FRONT VIEW

SECTION A

PLAN OF PROCEDURE

1 Adjust the size of the stand, if necessary.
As shown, the drill press stand is 24 inches high and holds a machine 40 inches tall. If your drill press is taller or shorter than this, you may want to adjust the height of the stand to keep it at a comfortable working height. Most craftsmen prefer to keep the drill press table at chest level (between 44 and 52 inches above the floor).

When you have determined the height of the stand you need, adjust the dimensions of the sides and back. You may also have to change the number and the height of the drawers from what's shown.

2 Select the stock and cut the parts to size.
To build the drill press stand as designed, you need

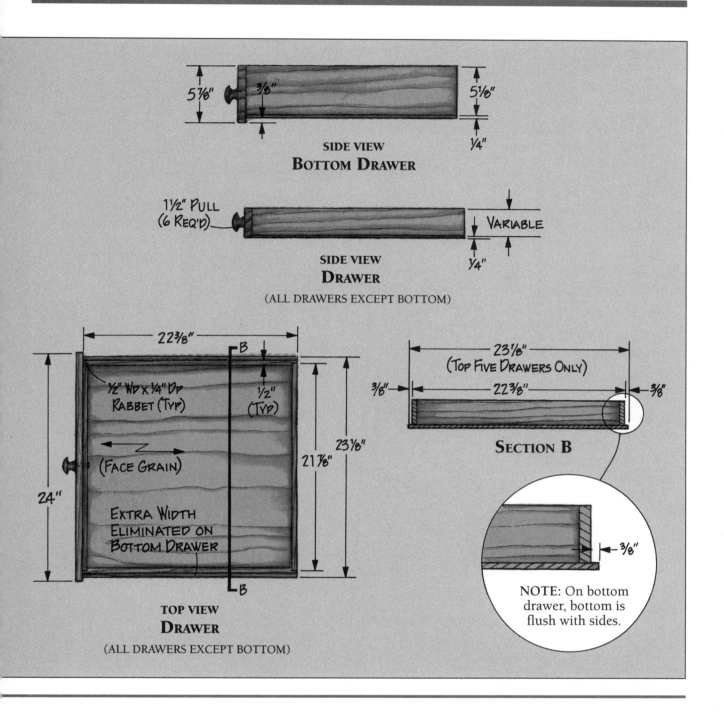

5⅞" ⅜" 5⅛" ¼"

SIDE VIEW
BOTTOM DRAWER

1½" PULL
(6 REQ'D) VARIABLE ¼"

SIDE VIEW
DRAWER
(ALL DRAWERS EXCEPT BOTTOM)

22⅜" B ½" (TYP)
½" WD x ¼" DP
RABBET (TYP)
(FACE GRAIN) 21⅞" 23⅛"
24"
EXTRA WIDTH
ELIMINATED ON
BOTTOM DRAWER
B

TOP VIEW
DRAWER
(ALL DRAWERS EXCEPT BOTTOM)

23⅛"
(TOP FIVE DRAWERS ONLY)
⅜" 22⅜" ⅜"

SECTION B

NOTE: On bottom
drawer, bottom is
flush with sides.

⅜"

approximately 15 board feet of 4/4 (four-quarters) hardwood lumber, one sheet of ¾-inch cabinet-grade plywood, and one sheet of ¼-inch plywood. The stand shown is made from maple lumber and birch plywood.

Cut the ¾-inch plywood parts to the lengths and widths needed. Plane the lumber to ¾ inch thick, and cut the valance, kickboard, braces, and drawer run-

ners to size. Plane the remaining lumber to ½ inch thick to make the drawer parts, but don't cut them yet; wait until after you have built the case.

3 Cut the joinery in the sides, top, and back.
The plywood parts of the case are assembled with

rabbets and dadoes. Lay out and cut the joinery, using a router or a dado cutter:

■ ¾-inch-wide, ⅜-inch-deep rabbets in the top edges of the sides to hold the top, as shown in the *Front View*

■ ¾-inch-wide, ⅜-inch-deep dadoes in the sides and back to hold the bottom

■ ⅜-inch-wide, ⅜-inch-deep rabbets in the side edges of the back, as shown in the *Top View,* and the top edge of the back, as shown in *Section A*

■ ⅜-inch-wide, ⅜-inch-deep dadoes in the back edges of the sides and top to hold the back

■ ⁵⁄₁₆-inch-wide, ⅜-inch-deep dadoes in the sides to hold the drawers

Note: Remember, the sides are *mirror images* of each other. Lay out and cut the joints on the right side so they are reversed from those on the left.

4 Drill holes in the sides. When you cut the drawer dadoes in the sides, they will probably cup toward the inside face. To counteract this tendency, you must install a steel rod between them, as shown in the *Front View.* To hold the ends of the rod, drill a ⅜-inch-diameter, ¼-inch-deep hole in each side, where shown in *Section A.*

5 Assemble the case. Finish sand the parts of the case. Assemble the sides, back, top, and bottom with glue. Glue the braces to the underside of the top, the drawer runners to the bottom, and the valance and kickboard to the front of the case. Reinforce the glue joints that hold the braces, valance, and kickboard with flathead wood screws, counterboring and countersinking the heads. Cover the screw heads with wooden plugs, and sand them flush with the surrounding surface.

Cut a ⅜-inch-diameter steel rod 23 inches long. Spread the sides apart temporarily, making them bow out, and put the rod in the stopped holes in the sides. Let the sides spring back into place — the rod will keep them straight.

6 Cut the drawer parts to size. No matter how carefully you work, the size of a case sometimes changes as you build it. For this reason, it's best to wait until after you've assembled the case before cutting the drawer parts.

Measure the width of the opening in the front of the case. If it varies from what is shown in the drawings, adjust the sizes of the drawer parts to compensate. Then cut the drawer parts to size. **Note:** The completed drawers should be ⅛ inch *narrower* than the opening to

work properly. Remember this when figuring the sizes.

7 Cut the drawer joinery. Like the case, the drawers are assembled with simple rabbets and dadoes. Cut ½-inch-wide, ¼-inch-deep rabbets in the front and back ends of the drawer sides, as shown in the *Drawer/Top View.*

8 Assemble and fit the drawers. Lightly sand the drawer parts. Nail and glue the drawer fronts, backs, sides, and bottoms together. Set the heads of the nails below the wood surface. Do not glue the drawer faces to the fronts yet. **Note:** The drawer bottoms on the top five drawers should protrude beyond the sides, as shown in *Section B.*

Slide the drawers into the case, fitting the protruding drawer bottoms into the ⁵⁄₁₆-inch-wide dadoes in the sides. If any of the drawers bind in the case, sand or scrape the drawer parts that rub until they slide in and out easily.

Tape ⅛-inch-thick shims to the outside of the backs of the drawers, near the corners. Slide the drawers all the way into the case — the shims should hold the drawer fronts flush with the front edges of the case. Place the drawer faces over the fronts, stacking them one on top of the other. Use scraps of cardboard or toothpicks as spacers to create ¹⁄₃₂- to ¹⁄₁₆-inch gaps between them. Temporarily attach the faces to the case with masking tape; then drill ⅛-inch-diameter pilot holes through the faces and the fronts where you will later attach drawer pulls.

Remove the drawer faces and discard the tape. Spread glue on the drawer fronts, and put the faces back in place. Temporarily drive #8 flathead wood screws through the pilot holes in the faces and into the fronts — this will hold the faces in place until you can apply the clamps. Remove the drawers from the case, and clamp the faces to the drawer fronts. Let the glue dry, then back the screws out of the pilot holes. Enlarge the holes and mount drawer pulls.

9 Finish the drill press stand. Remove the drawers from the case. Do any necessary finish sanding and apply several coats of a wiping oil such as tung oil, letting each coat dry thoroughly before applying another. Rub out the last coat with #0000 steel wool and paste wax — the wax will help the finish resist spills and other abuse. Also wax the drawer bottoms and the dadoes that hold them, to help the drawers slide smoothly. Mount the casters on the bottom of the stand, bolt the drill press to the top, and replace the drawers in the case.

8

Mission Occasional Table

fter decades of excessive Victorian ornament, the American Arts and Crafts movement was a breath of fresh air. The practitioners of this turn-of-the-century design philosophy advocated a return to simpler forms. "Beauty springs from the marriage of craftsmanship and utility . . . not affected ornament," wrote Gustav Stickley, one of the best-remembered makers of Arts and Crafts (or *Mission*) furniture. Stickley cut his woodworking teeth in a Shaker chair factory, where he learned the value of simple, functional furniture. Later, his respect for simplicity left its mark on many Mission designs.

This Mission-style occasional table is patterned after one of Stickley's better-known pieces, his "flower" table. (Originally, the top was shaped like the petals of a flower.) Every part is functional; every line is as simple as possible. The round top and four straight legs are held together by an X-shaped stretcher frame. And the table knocks down when it's not needed — just remove the four wedges or *tusks* from the stretcher frame and pull the legs off.

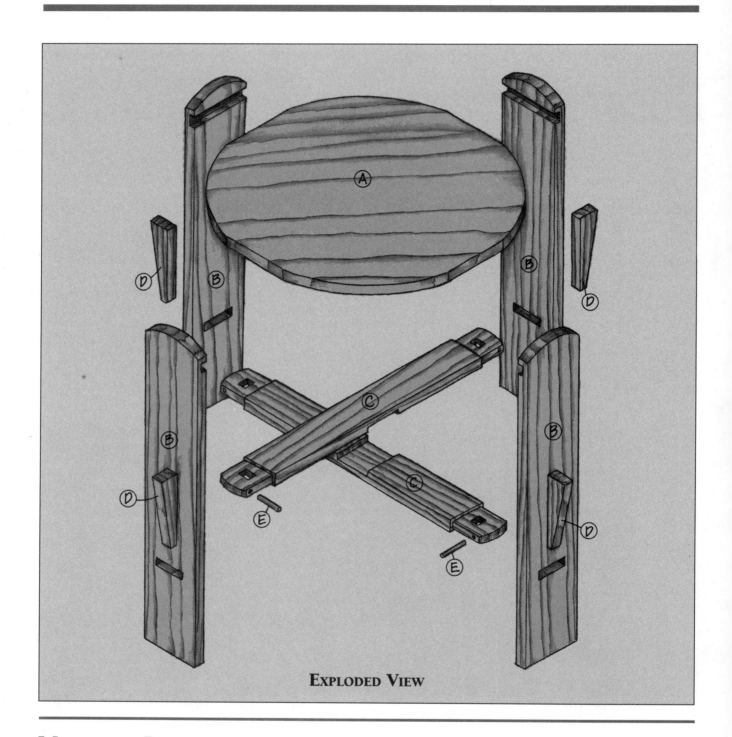

EXPLODED VIEW

MATERIALS LIST (FINISHED DIMENSIONS)

Parts

A. Top $\frac{3}{4}$" x 18" x 18"
B. Legs (4) $\frac{3}{4}$" x $4\frac{1}{4}$" x 22"
C. Stretchers (2) $\frac{3}{4}$" x $2\frac{1}{2}$" x $21\frac{3}{8}$"
D. Tusks (4) $\frac{3}{4}$" x $1\frac{5}{16}$" x 5"
E. Dowels (4) $\frac{1}{4}$" dia. x $2\frac{1}{4}$"

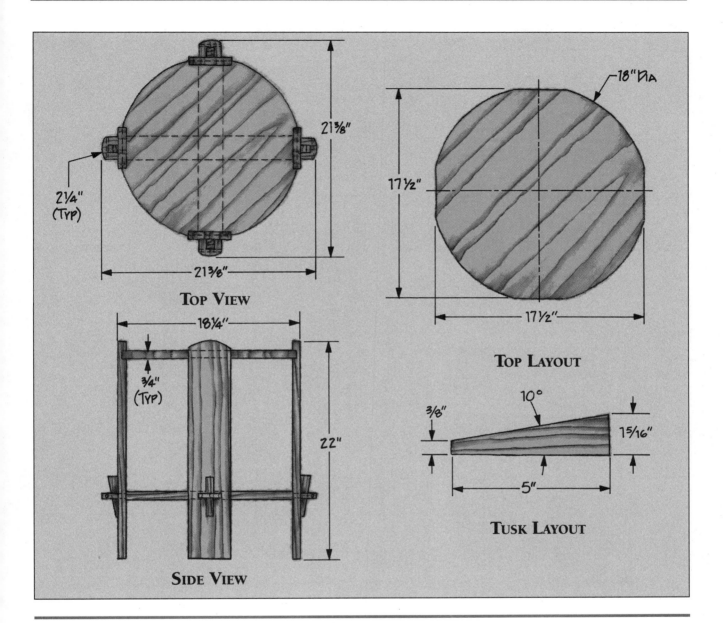

TOP VIEW

21⅜"

2¼"
(TYP)

21⅜"

18¼"

¾"
(TYP)

22"

SIDE VIEW

18" DIA

17½"

17½"

TOP LAYOUT

3⁄8"

10°

1⁵⁄₁₆"

5"

TUSK LAYOUT

PLAN OF PROCEDURE

1 Select the stock and cut the parts to size.
To make this table, you'll need about 8 board feet of
4/4 (four-quarters) lumber. You can use any cabinet-
grade wood, but Mission furniture was most often
made from oak or mahogany. The table shown is
made of white oak and stained with lye to give it a
weathered appearance.

Plane the lumber to ¾ inch thick. Glue up a wide
board to make the top, then cut the parts to the sizes
given in the Materials List. Note that the grain direction of the top must run *diagonally*. Taper one edge of
each tusk at 10 degrees, as shown in the *Tusk Layout*.

2 Cut the dadoes and lap joints. The stretcher
frame is assembled with a cross lap, and the tabletop
is held in dadoes. You can cut both of these joints
with a table-mounted router or a dado cutter:
■ a 2½-inch-wide, ⅜-inch-deep lap dado in each
stretcher, as shown in the *Stretcher Layout*
■ ¾-inch-wide, ⅜-inch-deep dadoes in the inside
faces of the legs, as shown in the *Leg Layout*

3 Make the mortises and tenons. The stretchers
hold the legs together with *tusk mortise-and-tenon
joints*. Tenons on the ends of the stretchers extend

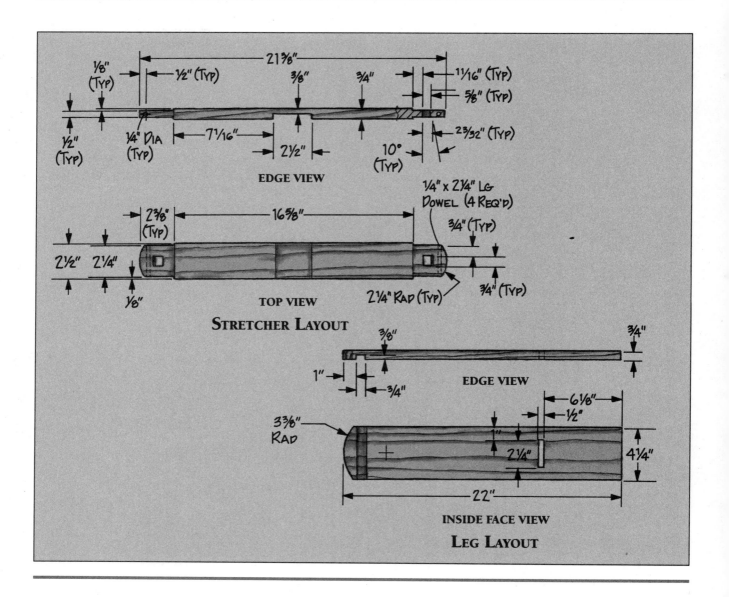

EDGE VIEW

STRETCHER LAYOUT

TOP VIEW

EDGE VIEW

INSIDE FACE VIEW

LEG LAYOUT

through mortises in the legs. These tenons are also mortised so they can be pinned in place with wedge-shaped tusks.

Make the leg or *primary* mortises first. Drill a line of overlapping ½-inch-diameter holes in each leg to create a slot ½ inch wide and 2¼ inches long. Then square the ends and straighten the sides of the slot with a chisel to make a square mortise, as shown in the *Leg Layout/Inside Face View.*

Cut the tenons in the stretchers to fit the mortises. Make four rabbets 2⅜ inches wide and ⅛ inch deep in each end of each stretcher, then shave the tenons down with a rabbet plane or a chisel until you get a slip fit.

Insert each stretcher tenon through its primary leg mortise, and mark a line on the tenon even with the

outside surface of the leg. Lay out the tusk mortise on the tenon so the mortise's *inside* edge is 1/16 inch inside the line. Drill out most of the waste from the mortises, and finish cutting them with a chisel. (*SEE FIGURE 8-1.*) **Note:** Because the stretchers are joined in the middle with a lap joint, you must make the tusk mortises on one stretcher with the lap dado facing *up,* and the mortises on the other stretcher with the lap dado *down.*

TRY THIS TRICK

Label the *inside* surfaces of the primary and tusk mortises so you'll remember which stretcher end fits each leg.

4 **Reinforce the tenons.** If you force the tusk down into its mortise — or if the tusk expands after it's seated in the mortise — there is a chance that the end of the tenon may split out. To prevent this, drill a ¼-inch-diameter hole horizontally through the end of each tenon, as shown in the *Stretcher Layout/Edge View.* Glue dowels in the holes, then sand or file the ends of the dowels flush with the edges of the tenons.

5 **Cut the shapes of the legs, tenons, and top.** Lay out the rounded edges and ends of the legs, stretchers, and top. Cut the shapes with a band saw or saber saw and sand the sawed edges.

6 **Assemble and finish the table.** Dry assemble (without glue) the parts of the table to make certain all the joints fit correctly. When you're satisfied they do, glue the stretchers together to make an X-shaped frame. Let the glue dry and finish sand all wooden surfaces. Apply a finish and rub it out with paste wax.

Note: The table shown is stained with lye to give it the old, weathered look of traditional Mission furniture. (Gustav Stickley used this chemical stain on occasion.) To make lye stain, mix 2 ounces of common household lye (sodium hydroxide, available in any grocery store) and 1 cup of water in a *glass* container — the lye would eat through metal. In another container, mix 1 heaping tablespoon of cornstarch into 2 cups of water. Slowly pour the cornstarch solution into the lye solution, stirring constantly. As you do so, the two liquids will thicken and make a paste. Apply the paste to the wood with a brush, let it work for 10 to 15 minutes, and wash it off with a weak solution of water and vinegar. (Mix 1 part vinegar with 3 parts water.) The vinegar will neutralize the lye and preserve the color. Let the surfaces dry thoroughly before you apply a finish over the stain.

A Safety Reminder

Wear a full face shield, rubber gloves, and a plastic apron when you apply the lye paste. Lye is caustic and can burn. If you get any paste on your skin, wash it off with water *immediately*.

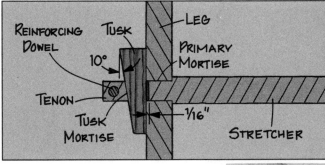

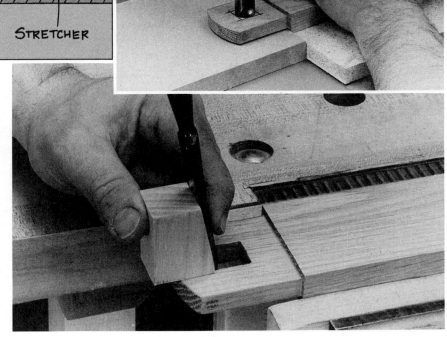

8-1 To make each tusk mortise, remove most of the waste with a ⅝-inch drill bit, then square the corners and straighten the sides with a chisel. Note that the *outside* face of the mortise is angled to match the tusk. To cut this angle accurately, taper one edge of a small hardwood block at 10 degrees. Secure the block to the tenon with double-faced carpet tape and use it to guide the back of the chisel. When the tusk mortise-and-tenon joint is assembled, a small part of the tusk mortise should rest just inside the primary mortise. This design lets the tusk hold the leg tight against the shoulders of the tenon.

ROUND TUSK MORTISE-AND-TENON JOINTS

Tusks and tusk mortises don't have to be square. You can cut a round tusk from a dowel and drill a round mortise to fit it.

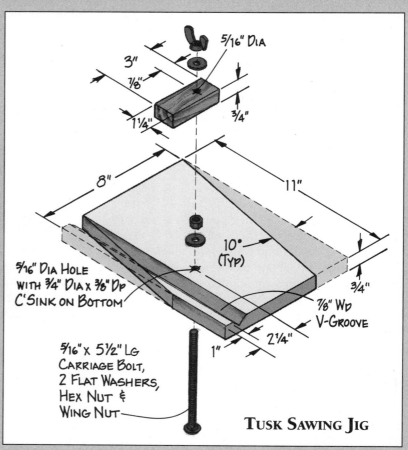

TUSK SAWING JIG

1 **To make a round tusk, slice a** 1-inch-diameter dowel at a 10-degree angle, as shown. Rest the dowel in a V-jig to hold it as you cut. Sand the sawed surface on a belt sander or disc sander to flatten it.

2 **Drill a 1-inch-diameter round** mortise in the tenon at a 10-degree angle, matching both the size and the taper of the tusk. As with square tusks, a small portion of the round tusk mortise should rest inside the primary mortise when the joint is assembled. Additionally, the tusk mortise must be angled toward the *inside* of the joint. To assemble the joint, insert the tenon through the primary mortise. Drop the round tusk into the tusk mortise so the flat, tapered surface will rest against the wood, holding the two parts together.

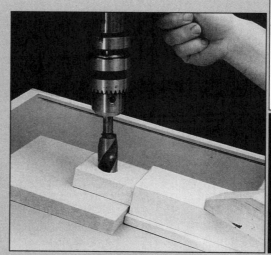

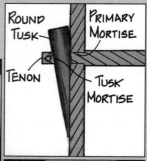

9

WOODEN PENS AND PENCILS

With a simple hardware kit, you can make an elegant pen or pencil from almost any wood. Just drill a hole in two small wooden blocks, turn them round, and press the hardware into the turned cylinders. Furthermore, you don't need a lathe to turn the wooden parts — you can make the entire project on a drill press! Use the turning jig shown in "Turning" on page 89 to create the cylinders.

PLAN OF PROCEDURE

1 Select the stock and cut the parts to size. You can make pens and pencils from almost any wood, although hardwood works best. Choose a wood with distinctive coloring or grain patterns, such as bird's-eye maple, walnut burl, or ash crotch — these make the completed writing instruments visually interesting. The pens shown are made from *bocote,* an imported hardwood.

Cut two turning blocks from the stock. Make these blocks the same length as the brass tubes that come with the hardware kit, and ¼ inch wider and thicker than their diameter. For most of the available kits, the blocks should be about 2¹⁄₁₆ inches long and ⅝ inch square.

2 Insert the tubes in the turning blocks. Drill a 7-millimeter-diameter hole through the length of each turning block. *(SEE FIGURE 9-1.)* Then glue the brass tubes in the holes with cyanoacrylate glue (Super Glue). *(SEE FIGURE 9-2.)*

3 Turn the cylinders. When the glue dries, mount the turning block on a mandrel and secure the mandrel in your drill press chuck. Remove the tailstock from the turning jig, and clamp the jig to the drill press table. Using the tool rest to support a chisel, turn the block round. Repeat for the remaining block. *(SEE FIGURE 9-3.)* Sand and finish the wooden cylinders while they are mounted on the drill press. *(SEE FIGURE 9-4.)*

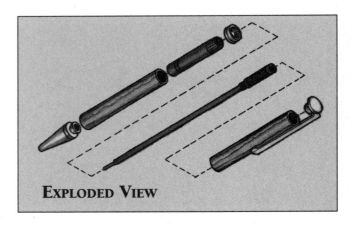

EXPLODED VIEW

4 Press the metal parts in the cylinders. Put three layers of masking tape on the bottom of the chuck, to prevent it from marring the brass pen parts. Following the directions that come with the hardware, press the metal parts into the brass tubes. Start each part by tapping it with a rawhide mallet, then use the quill as a press to gently push it all the way into the tube. *(SEE FIGURE 9-5.)*

WHERE TO FIND IT

You can purchase pen and pencil hardware kits, 7-millimeter drill bits, mandrels, and collars from:

Woodcraft Supply Corp.
210 Wood County Industrial Park
Parkersburg, WV 26102

Woodworker's Supply
1108 North Glenn Road
Casper, WY 82601

9-1 To drill a hole through the turning block, clamp it in a vise or a hand screw clamp. Press the edge or the face against the bit to make sure the length of the block is parallel to the bit. Then drill a 7-millimeter-diameter hole through the block.

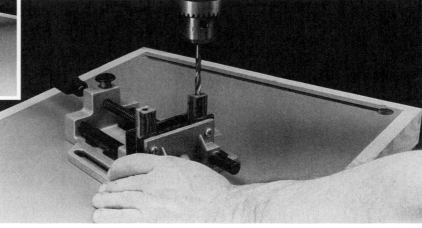

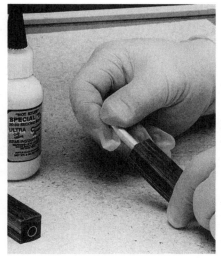

9-2 Coat the outside of the brass tube with cyanoacrylate glue (Super Glue) and insert the tube in the block. As you push it in, rotate the tube, if possible — this will help spread the glue. Press the tube in until the ends are flush with the ends of the block. **Warning:** Wear rubber gloves as you do this to keep from getting glue on your skin.

9-3 Let the glue dry completely, then mount the turning block on a mandrel. Secure the mandrel in the drill press chuck. Remove the tail-stock from the turning jig, clamp the jig to the table, and adjust the tool rest to within ⅛ inch of the block. Make sure the block doesn't hit the rest as it rotates. Then start the drill press and turn the block to a cylinder. Use a gouge to rough out the turning, then turn it down to its final diameter with a flat chisel. **Note:** Most pen-making mandrels come with brass collars that eliminate the need to check your work with calipers. Just turn the cylinder until it's the same diameter as the collars (usually 8.5 millimeters).

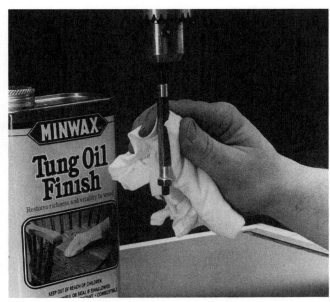

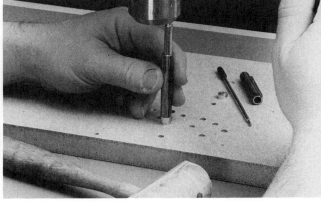

9-4 When the cylinder is the proper diameter, sand it smooth on the mandrel. Then apply a finish. The choice of finish depends on the wood and your own preference. A wiping oil (such as tung oil) works well for most woods. For oily woods, simply apply paste wax.

9-5 Use the quill and the quill feed as a press to install the metal parts in the turned cylinders. Start them by tapping with a small rawhide mallet. To protect the brass pen parts, apply three layers of masking tape to the bottom of the chuck. Then gently push them home with the quill, making sure the parts slide *straight* into the brass tubes. The order in which you install the hardware will depend on the manufacturer and whether you're making a pen or a pencil — check the instructions.

INDEX

Note: Page references in *italic* indicate photographs or illustrations.
Boldface references indicate charts or tables.

WOODWORKING GLOSSARY

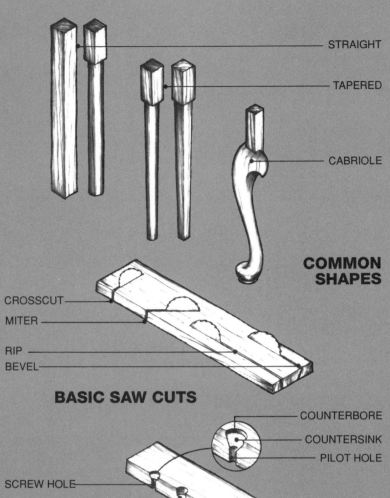

COMMON SHAPES

- STRAIGHT
- TAPERED
- CABRIOLE

BASIC SAW CUTS

- CROSSCUT
- MITER
- RIP
- BEVEL

HOLES

- COUNTERBORE
- COUNTERSINK
- PILOT HOLE
- SCREW HOLE
- STOPPED HOLE
- THRU HOLE

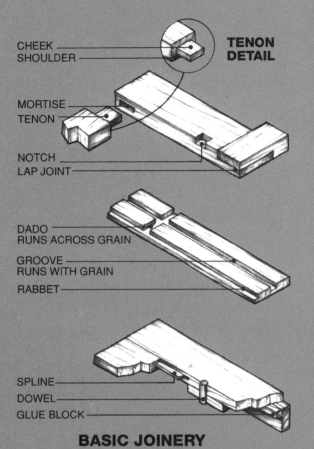

TENON DETAIL
- CHEEK
- SHOULDER

- MORTISE
- TENON
- NOTCH
- LAP JOINT

- DADO RUNS ACROSS GRAIN
- GROOVE RUNS WITH GRAIN
- RABBET

BASIC JOINERY
- SPLINE
- DOWEL
- GLUE BLOCK

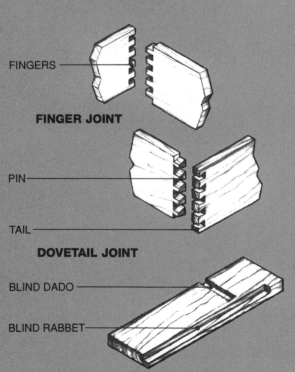

FINGER JOINT
- FINGERS

DOVETAIL JOINT
- PIN
- TAIL

SPECIAL JOINERY
- BLIND DADO
- BLIND RABBET

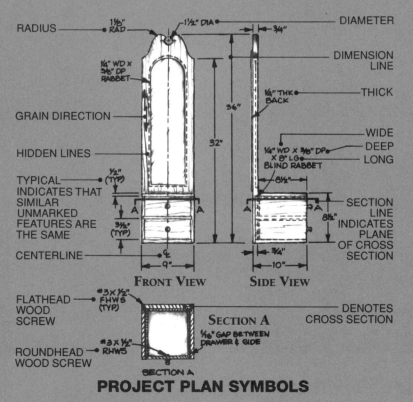

PROJECT PLAN SYMBOLS

- RADIUS — 1⅛" RAD
- 1½" DIA
- ¾"
- DIAMETER
- ¼" WD x ⅜" DP RABBET
- DIMENSION LINE
- GRAIN DIRECTION
- ¼" THK BACK — THICK
- HIDDEN LINES
- ¼" WD x ⅜" DP X 8" LG BLIND RABBET — WIDE / DEEP / LONG
- 36"
- 32"
- TYPICAL INDICATES THAT SIMILAR UNMARKED FEATURES ARE THE SAME — ½" (TYP)
- 3½" (TYP)
- 8½"
- 8½"
- SECTION LINE INDICATES PLANE OF CROSS SECTION
- CENTERLINE
- 9"
- ¾"
- 10"

FRONT VIEW   **SIDE VIEW**

- FLATHEAD WOOD SCREW — #3 X ½" FHWS (TYP)
- ROUNDHEAD WOOD SCREW — #3 X ½" RHWS

SECTION A
- ⅛" GAP BETWEEN DRAWER & GIDE
- DENOTES CROSS SECTION

SECTION A